OAKWOOD LIBRARY OF RAILWAY HISTORY

THE SEVERN BRIDGE RAILWAY

COLIN G. MAGGS

Severn Bridge view west. Gauge huts are on each side of the line in the foreground, while the brake van stands at the end of the stone viaduct, 1956. *BR WR*

THE OAKWOOD PRESS

First published in the United Kingdom, 2023,
by The Oakwood Press,
54-58 Mill Square,
Catrine,KA5 6RD

01290 551122
www.stenlake.co.uk

ISBN 978-0-85361-769-3

Printed by
Claro Print, Office 26, 27, 1 Spiersbridge Way,
Thornliebank, Glasgow G46 8NG

Acknowledgements

Thanks for help are due to John Mann, Andrew Perrin, Alan Price and Marcus Sealy, with special thanks to Colin Roberts for checking and improving the text.

Note to the Reader

Train times in this book are those given in official timetables, the 12-hour clock being used before June 1963 and the 24-hour clock subsequently. To avoid confusion, 'Up' and 'Down' are capitalised when used for direction, with lower case for gradients.

'2251' class 0-6-0 No. 2251 leaving Lydney for Bristol; the brake van bears the legend 'Bristol West Depot RU Not in Common Use No 2 Transfer'. 'RU' means 'restricted user' meaning the vehicle could only be used for the service designated. *Revd. Alan Newman*

Contents

'16XX' class 0-6-0PT No. 1639 leaves Severn Bridge station with an Up train, 19th July, 1958. *R. E. Toop*

The Severn Bridge with the Gloucester & Berkeley Canal in the foreground, 1949.
Author's collection

Midland Railway map of the Berkeley Road – Lydney line.

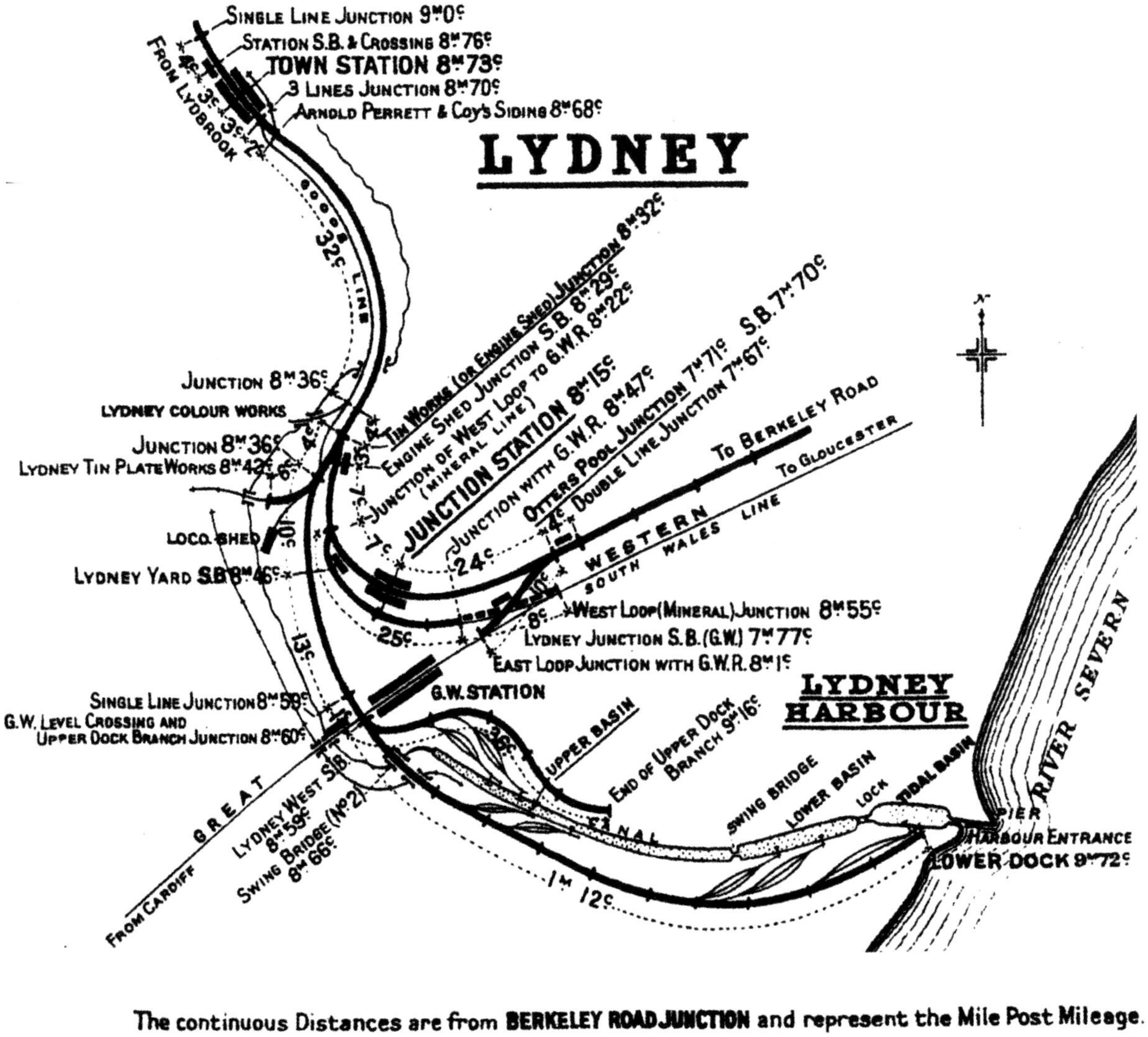

Introduction

What are the longest railway bridges in Great Britain? Most people would rightly name the Tay and Forth Bridges. A few might remember the third, the ill-fated Solway Viaduct which linked England with Scotland. And then what about the fourth, the longest railway bridge wholly in England?

The answer is the Severn Bridge between Sharpness and Lydney. It was relatively unknown because it was normally used by just local trains carrying a dozen or so passengers and did not normally see important passenger trains except on winter Sundays when the Severn Tunnel was closed for maintenance and the bridge used as a diversionary route. Passengers were then vexed at the longer time spent on the journey, instead of realising that they were being granted a bonus by being given a ride across a principal engineering structure.

The bridge could be seen by passengers travelling on the South Wales Railway between Gloucester and Chepstow, but if travelling by road it was not readily visible and a special diversion was necessary.

The Severn Bridge Railway Company had a run of bad luck. It had the potential of becoming a through route and as such would have been a gold mine. Unfortunately events outside the company's control – direct routes to London and Southampton planned by other companies failing; coal and tin strikes in the Forest of Dean reducing the need for transport, coupled with ships increasing in size thus rendering the coal tip at Sharpness unusable, meant that the line lacked profitability.

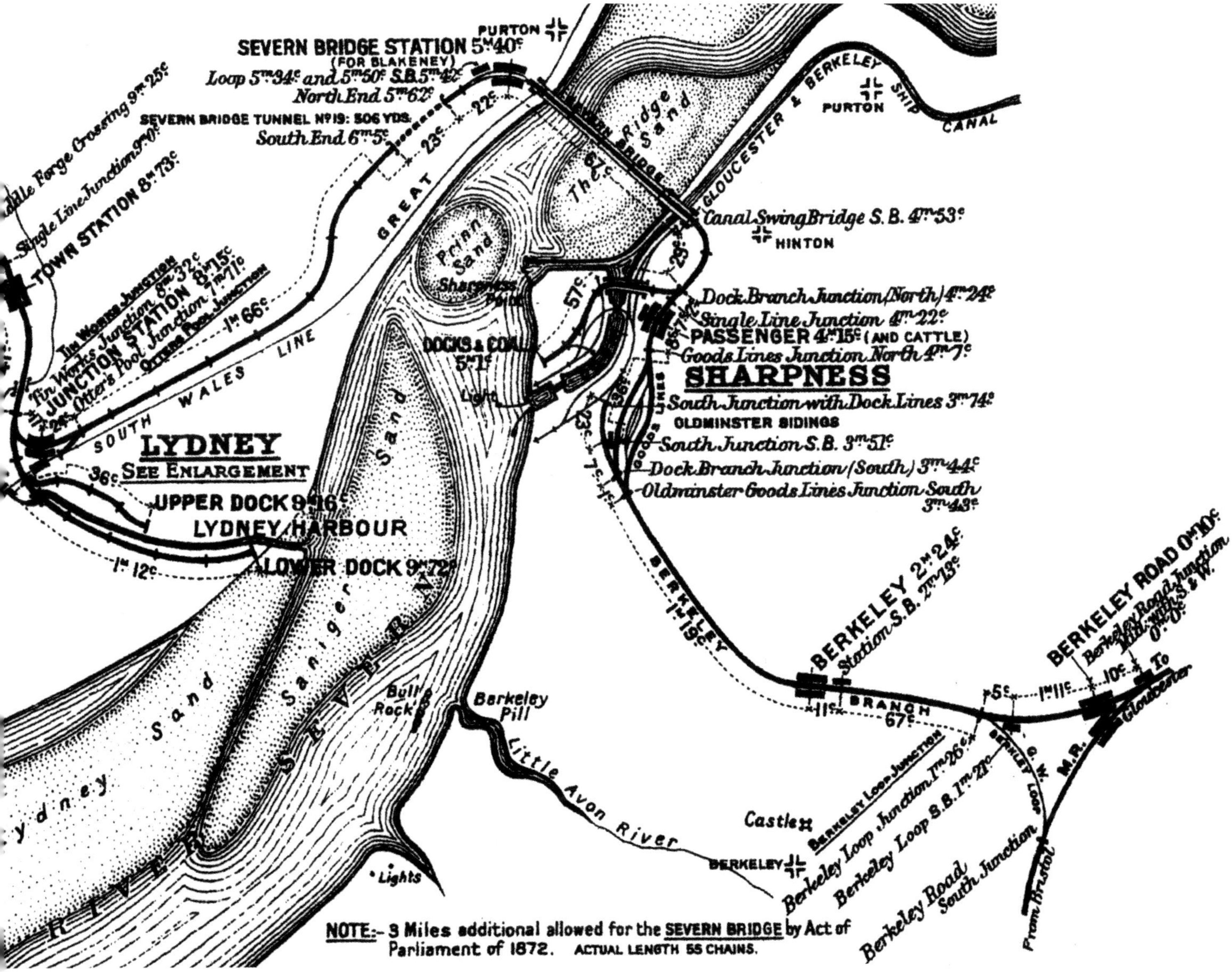

The continuous Distances are from BERKELEY ROAD JUNCTION and represent the Mile Post Mileage.

Severn Bridge view east, 26th April, 1949. *BR WR*

Chapter One

The Line's Advent

There was no land crossing of the Severn below Gloucester before the 19th century. About 1810 a tunnel under the Severn at Newnham was promoted and a shaft sunk midway between Newnham church and Bullo Pill. Excavation below the river was well under way when water broke in and workmen had a narrow escape from drowning. The subsequent investigation showed that the damage was irreparable so the undertaking was abandoned.

The Bristol & Gloucester Railway along the eastern bank of the Severn opened on 8th July, 1844 linking Gloucester with Bristol and Bath, and the South Wales Railway on the western bank was completed on 19th July, 1852 linking Gloucester with Newport and Cardiff.

Bristol and Bath on one shore of the Severn had a poor connection with Newport and Cardiff on the other side. A rail journey from Bristol to Cardiff entailed travel via Gloucester, a distance of 93½ miles, whereas a crow could make the journey in just 24 miles: it was obvious that a short cut was needed.

In 1844 Isambard Kingdom Brunel proposed the South Wales & South of Ireland Railway, a broad gauge line running from the Cheltenham & Great Western Union Railway at Stonehouse, crossing the Severn between Fretherne and Awre by two timber bridges, one 780 yards in length and the other 800 yards. The Severn Navigation Commissioners accepted the proposal, but the Admiralty vetoed the bridge, thereby putting a stop to the scheme. The reason is difficult to understand as most shipping used the Gloucester & Berkeley Ship Canal (also known as the Gloucester and Sharpness Canal) which had opened on 26th April, 1827 and provided a much easier, safer and shorter route for waterborne traffic to Gloucester.

In 1845 the South Wales Railway proposed a Bill for an 858 yards long bridge 200 yards upstream from that proposed in 1844, also depositing plans for an alternative scheme using a tunnel should those for the bridge be rejected. The Admiralty stated that it would never sanction a bridge and the House of Commons Committee considered evidence in favour of a tunnel insufficient to warrant its approval.

Map of proposed start and end points for Severn crossings.

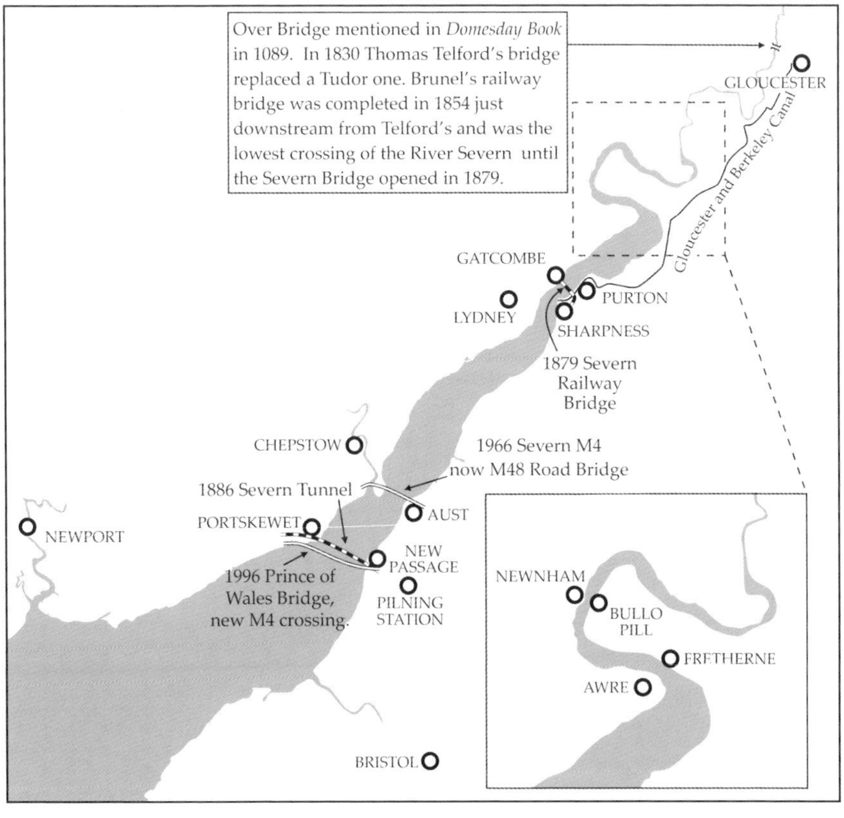

Simultaneously with the publication of the Parliamentary notice of the second South Wales Railway scheme, in November 1845 an application was made to Parliament for a high-level bridge across the Severn at Aust. At an elevation of 150 feet above high water it was to be partly of stone, but with the two central arches of suspension design using iron. The scheme failed in its initial stage.

A temporary stop-gap was opened on 8th September 1863. This was the Bristol & South Wales Union Railway, comprising a line from Bristol to New Passage, and a ferry across to Portskewett with a branch line link to the South Wales main line.

With the technology of the age, the lowest feasible bridging point was from Sharpness across to Lydney, this route also having the advantage of being adjacent to the Forest of Dean coalfield, enabling fuel to be carried directly to the industrial Stroud Valley and other consumers.

In 1865 W. B. Clegram, engineer to the Gloucester & Berkeley Canal, employed George William Keeling of Lydney to explore the possibility of spanning the Severn near Sharpness. He found that the deep water channel remained in the same place indicating that shipping could keep to the same arches. George Wells Owen of Westminster, engineer to minor Forest railways, joined him in developing the idea; plans were deposited in 1870 but failed through lack of financial support.

With the increasing size of vessels, the lock at the Sharpness entrance to the Gloucester & Berkeley Canal was too small so some ships had to discharge into lighters. To solve the problem, W. B. Clegram planned a new entrance half a mile downstream having a larger lock and giving access to a new floating dock. An Act for carrying out this work was passed in 1870, construction began the following year and it was officially opened 25th November, 1874.

Imports at Sharpness exceeded exports annually by 250,000 tons. Hitherto vessels bringing corn, timber and so on, after discharging at Sharpness, took on ballast and proceeded to Cardiff or another Bristol Channel port for a return cargo, usually coal or iron. The construction of a bridge would allow the ships to be loaded with South Wales steam coal, or whatever, at Sharpness, thus saving the costs of loading and unloading the ballast, the time taken for a trip to another port and having to pay another set of pilotage dues.

As the demand for Forest coal was for household, not steam use, its market lay either to the east or in Somerset and Devon. For the latter destinations, it was carried by water to Bridgwater or Highbridge and then onwards by rail. Forest coal had a brittle nature and deteriorated when moved: such as from mine to the Severn & Wye Railway; the Severn & Wye to ship; then ship to rail, whereas a Severn Bridge would allow it to be transported direct from mine to coal merchant. A bridge would also favour the South Wales steam coal market.

1865 saw no less than four schemes projected:

1. A tunnel crossing the Severn below Chepstow with two air shafts midway. As these were considered objectionable, its engineer, Charles Richardson who was later to be responsible for the tunnel which was eventually cut, modified the design placing the shafts each side of the river.
2. A high-level bridge near Chepstow.

3+4 A bridge across the Severn near Newnham.

The high-level bridge was regarded with the greatest disfavour due to its impediment to navigation and the number of piers favouring the accumulation of sand, while the other bridges were opposed for their navigational hazards. Ships of about 400 tons register would have to lower the top gallant masts and this inconvenience and delay could cause them to miss a tide. Two Bills, the high-level bridge and the Severn Junction Stonehouse-Newnham, came before the House of Commons Committee on 16th May, 1865 and although strenuously opposed, were passed on 26th May, then on 30th June passed the Committee of the House of Lords and received Royal Assent on 5th July, 1865, *28 & 29 Vic cap 366*. This Act empowered raising £420,000 in shares and £140,000 on loan. Both schemes collapsed through lack of finance.

The idea persisted and in November 1865 Parliament was given notice of two schemes crossing between Gatcombe and Purton. In January 1866 these were referred to the engineers of the Midland Railway, Great Western Railway and the South Wales Railway which resulted in both plans being abandoned.

In 1870 two projects were put forward: a tunnel at Portskewett, and at Purton a road and a dual (broad/standard) gauge railway bridge, but both schemes collapsed due to lack of money.

1871 saw six schemes projected, and as three were simply entitled the Severn Bridge Railway, for Parliamentary purposes they were numbered:

1. To run from the South Wales Railway east of Chepstow Bridge to the Bristol & South Wales Union Railway east of Pilning station with a branch to Thornbury. It was planned to cross the river by a viaduct 1,780 yards long with 27 spans, the two largest of 800 feet span. This Bill failed to comply with standing orders.
2. A rail and footpath bridge across at Sharpness. The bridge 1,404 yards in length, was to consist of 28 spans, the largest of 300 feet span and 69 feet high. The design with 28 spans was criticised by the Severn Commissioners who had concerns over the navigation channel becoming silted, so the design was amended to 21 spans with larger openings.

This latter scheme was eagerly backed by the Gloucester & Berkeley Canal which was currently constructing large docks at Sharpness, and the plan also keenly supported by the Severn & Wye Railway and the Midland Railway. The latter, together with the canal company, guaranteed 4½ per cent on a £75,000 debenture loan. The GWR, expecting to lose a certain amount of traffic if the line was built, opposed the bill.

3. A line from the South Wales Railway west of Portskewett across a viaduct to the Bristol & South Wales Union Railway near New Passage. The viaduct was to be 3,920 yards in length with 71 spans, the largest being 700 feet. This Bill was abandoned.

IN PARLIAMENT—SESSION 1872.

THE SEVERN BRIDGE RAILWAY.

(Incorporation of Company—Construction of Railway from the South Wales Railway, near Chepstow Bridge, to the Bristol and South Wales Union Railway, and of Railways to connect intended Railway with the Wye Valley Railway and the Midland Railway—Working arrangements with Great Western and Midland Railway Companies—Amendment of Acts).

Notice is hereby given, that it is intended to apply to Parliament, in the ensuing Session, for leave to bring in a Bill for the following or some of the following purposes:—

To incorporate a Company herein referred to as the Company, and to enable them to make and maintain all or some one or more of the following Railways, or some part or parts thereof respectively, with all needful works, stations, and approaches, and conveniences connected therewith, that is to say:—

A Railway (No. 1) commencing by a junction with the South Wales Railway of the Great Western Railway Company, at a point one chain or thereabouts from the easternmost part of the Chepstow Bridge, carrying the said Railway over the River Wye, in an Enclosure numbered 333 on the Tithe Commutation Map of the Parish of Tidenham, in the County of Gloucester, passing over the River Severn by a Bridge, and terminating by a Junction with the Bristol and South Wales Union Railway of the Great Western Railway Company, in the Parish of Almondsbury, at a point ten chains or thereabouts eastward of the post denoting nine miles on that Railway, and which said intended Railway will be situate in the parishes or places following, or some of them, that is to say:—Sedbury, Beachley, Tidenham, Aust, Northwick, Henbury, Elberton, Westbury, Olveston, and Almondsbury, all in the County of Gloucester.

A Railway (No. 2) wholly situate in the Parish of Tidenham, in the County of Gloucester, commencing by a Junction with the authorised Wye Valley Railway (No. 1) at a point 4 furlongs and 6 chains from the commencement thereof, which said Junction will be in an Enclosure numbered 3:2 on the Tithe Commutation Map of the said Parish of Tidenham, and terminating by a Junction with the before-mentioned intended Railway (No. 1), in an Enclosure numbered 195 on the Tithe Commutation Map of the said Parish.

A Railway (No. 3) wholly situate in the said Parish of Tidenham), commencing by a Junction with the South Wales Railway of the Great Western Railway Company, at a point 13 chains and 15 yards measuring along the said Railway in an easterly direction from the post denoting 140½ miles on that Railway, and which Junction will be in an Enclosure numbered 319 on the Tithe Commutation Map of the said Parish of Tidenham, and terminating by a Junction with the before-mentioned intended Railway (No. 1), in an Enclosure numbered 195 on the said Tithe Map.

A Railway (No. 4) commencing by a Junction with the aforesaid Railway (No 1), in an Enclosure numbered 2286 on the Tithe Commutation Map of the Parish of Henbury, in the County of Gloucester, and terminating by a Junction with the Thornbury branch of the Midland Railway Company, in an Enclosure numbered 406 on the Tithe Commutation Map of the Parish of Thornbury, at a point on the last-named Railway 27 chains or thereabouts from the Turnpike-road leading from Thornbury to Bristol, measuring in an easterly direction along that Railway, and which said intended Railway will be situate in the parishes or places following, or some of them, that is to say:—Henbury, Cote Farm, Elberton, Littleton-on-Severn, Thornbury, and Oldbury, all in the County of Gloucester.

To cross, divert, alter, or stop up, whether temporarily or permanently, roads, railways, drains, sewers, pipes, rivers, streams, and watercourses so far as it may be necessary in constructing or maintaining the said intended Railways and Works, to deviate from the lines of railway, both vertically and horizontally, to such an extent as may be defined in the Bill.

To purchase lands, houses, easements, and other property, compulsorily, and by agreement, for the purposes of the said intended Railways and Works, to levy tolls, rates, and charges in respect thereof, and to exercise other rights and privileges.

To enable the Company, on the one hand, and the Great Western Railway Company and Midland Railway Company, or either of those Companies on the other hand, from time to time, to enter into agreements with respect to the working, use, management, construction, and maintenance of the said intended Railways, or any of them, or any part or parts thereof respectively, the supply of Rolling Stock and Machinery, and of officers and servants for the conduct of the traffic of the Railways, the payment to be made and the conditions to be performed with respect to such working, use, management, construction, and maintenance, the interchange, accommodation, and conveyance of traffic coming from or destined for the respective undertakings of the contracting Companies, and the division and appropriation of the revenue arising from that traffic, and to authorise the appointment of joint committees for carrying into effect any such Agreement as aforesaid, and to confirm any Agreement which previously to the passing of the Bill may be made touching any of the matters aforesaid.

The Bill will vary and extinguish all existing rights and privileges which would interfere with its objects, and it will incorporate with itself the necessary provisions of "The Companies Clauses Consolidation Act, 1845," "The Companies Clauses Act, 1863," "The Companies Clauses Act, 1869," "The Lands Clauses Consolidation Acts, 1845, 1860, and 1869," "The Railways Clauses Consolidation Act, 1845," and "The Railways Clauses Act, 1863," and it will or may amend and enlarge the powers and provisions of the several Acts relating respectively to the Great Western and Midland Railway Companies.

Duplicate Plans and Sections, describing the lines, situation, and levels of the proposed Railways, and the lands, houses, and other property in or through which they will be made, or which may be taken under the powers of the Bill, together with a Book of Reference to such plans, containing the names of the owners or reputed owners and lessees, and of the occupiers of such lands and houses, and other property, also an Ordnance Map, with the lines of Railway delineated thereon, so as to show their general course and direction, and a copy of this Notice, will, on or before the 30th day of November instant, be deposited for public inspection with the Clerk of the Peace for the County of Gloucester, at his office at Gloucester, and on or before the same day a copy of so much of the said plans, sections, and Book of Reference as relates to each parish in or through which the intended Railways will be made, or in which any lands, houses or other property, are intended to be taken, and a copy of this Notice, will be deposited with the Parish Clerk of each such parish, at his Residence. And, in the case of any extra-parochial place, with the Clerk of some parish immediately adjoining such extra-parochial place, at his place of abode.

Printed Copies of the proposed Bill will be deposited in the Private Bill Office of the House of Commons on or before the 21st day of December next.

Dated this 13th day of November, 1871.

SUTTON and OMMANNEY,
80, Coleman-street, London,
Solicitors for the Bill.

7411

Notice of application to bring in the Severn Bridge Railway Bill, *Bristol Times & Mirror* 17th November, 1871.

The Severn Bridge Railway Bill No. 2 came before the Examiners on 7th February, 1872 and they reported that the standing order had been complied with. It then came before the Select Committee on 11th April, 1872, the first witnesses speaking of the line's advantages to both the Forest and Crown. Resolution in favour of the bridge had been passed by the Gloucester Chamber of Commerce and Gloucester Corporation was also in its favour.

On 12th April, 1872 the presiding officer of the Select Committee Sir Hadworth Williamson, asked W. P. Price, chairman of the Midland Railway in view of the failure of the 1870 scheme whether he could guarantee that sufficient capital could be raised. Price said that the MR and GWR could not subscribe without each other's consent and the GWR had raised last minute objections to the scheme. Representatives of the Severn & Wye Railway & Canal Company gave evidence as did owners of collieries and iron ore fields. The chairman said that he was convinced that the bridge would benefit Forest trade. Several Severn pilots offered evidence to prove that the bridge would not be an impediment to navigation.

The urgent necessity for a connection was tersely put to the Parliamentary Committee by witness Sir James Campbell who said:

> At present the eastern division of the county [of Gloucestershire] is separated from the western by the river Severn in such a way that though the people on either side can see each other, they are obliged to go round 30 miles in order to shake hands.

On 15th April, 1872 the engineer G. W. Keeling said that in his experience extending over 12 years there had been no variation of the channel where the bridge would cross and that an Admiralty chart of 1814 showed that there had been no alteration since that year. Thomas Elliott Harrison, consulting engineer to the Canal Company, gave evidence for the scheme's practicality and its estimate of £275,000.

On 16th April, 1872 ships' masters and pilots gave evidence against vessels passing through the bridge safely. Lord Fitzhardinge, owner of the Berkeley Castle estate which extended above and below the site of the proposed bridge, said that some of his land was 10 feet below the level of the regular spring tides and protected by an embankment, but that if any obstruction was placed on Wheel Rock, (where it was proposed to build a bridge pier), the rock would act as a breakwater, deflect the tide and have an injurious effect on his embankment. Spring tides rose to a height of 30 to 34 feet and travelled at speeds of 8 to 10 knots dependent on the direction and velocity of the wind. Where the bridge crossed the deep navigation channel to Wheel Rock, huge eddies and vortexes swirled.

Lord Fitzhardinge also owned a salmon fishery near the bridge which he let at £40 per annum and believed that the construction of a bridge would drive fish from that part of the river. G. W. Keeling being recalled, said that the reclaimed land on Lord Fitzhardinge's estate was nearly on a level with an ordinary spring tide and a high spring tide 3 to 4 feet above the land.

On 7th April, 1872 Handel Cosham, a Bristol colliery owner, said that the MR's branch from Berkeley Road to Sharpness would afford ample facilities for taking coal to the town so the bridge was unnecessary. Henry Hetheridge attached to the Royal School of Mines, said there were large quantities of coal in the Bristol field equally as good as Forest coal and that Bristol coal would be cheaper in the long run. A civil engineer, James Abernethy, said that assuming a bridge was necessary, a worse site for it could not have been selected in any part of the river. James Grierson, general manger of the GWR said that there was no need for the bridge as in a few months when the gauge conversion had been completed, the

GWR would be able to convey Forest minerals from Lydney to any part of the country on its own system.

On 18th April, 1872 summaries of evidence for and against were given and the chairman, Sir Hadworth Williamson, said that the committee would report that the preamble was proved, but it was of the opinion that clauses should be inserted giving Lord Fitzhardinge compensation for any injury to his property which might be sustained by the construction of a bridge; also compensation for the loss of his salmon fishery. The Bill received little opposition in the House of Lords and it received Royal Assent on 18th July, 1872, *35 & 36 Vic cap 109*.

The Severn Bridge Railway Company (SBR) was incorporated on 18th July, 1872 to build a line 4 miles 1 furlong in length to link the Great Western and the Severn & Wye Railways at Lydney with the branch of the Midland Railway at Sharpness. (Although a foot toll-bridge with a width of 6 feet was authorised, this was never constructed; it was stipulated that the fee for its use must be under three pence. The footbridge would have cost an additional £10,000. Gloucestershire County Council refused to advance the money, reasoning that if people wanted to reach the other side, they would go by train rather than face a windy walk of 1,387 yards).

The Severn & Wye Railway & Canal Company with which the SBR linked,was one of the oldest in the United Kingdom. The Lydney & Lydbrook Railway had been incorporated in 1809 and changed its name to the Severn & Wye Railway & Canal Company in 1810. A canal at Lydney and several colliery tramways were constructed in the Forest of Dean, but until 1853 only used privately until that year when the company obtained Parliamentary powers to become a public carrier. By 1864 it owned about 30 miles of 3 feet 8 inch gauge horse-worked tramroads and by the end of 1865 its engineer, G. W. Keeling had purchased five small tank engines with flangeless wheels for working the tramroad.

In order to suit its GWR neighbour, in 1867 it decided to have some of its track changed to broad gauge. By April 1869 three locomotives had their gauge altered and subsequently two new broad gauge engines were purchased. By 1872 the directors resolved to adopt standard gauge, so that year the three original narrow gauge and the two new broad gauge engines were suitably converted.

The total length of the Severn & Wye Railway was 38 miles 22 chains, but only 27 miles 45 chains were used for passenger and goods traffic, the rest being purely mineral line.

The authorised capital of the Severn Bridge Railway was £300,000, £225,000 in shares (£25,000 had been subscribed before the Act was passed) and £75,000 loans, while to offset the high cost for its length, the company was allowed to charge for an extra three miles over the bridge rather than the actual distance of less than a mile. The GWR, MR, Severn & Wye and the Gloucester & Berkeley Canal Companies were authorised to subscribe by this and subsequent Acts. The estimated cost of construction was £277,973.

Canals and the rivers Severn and Avon had been bringing to Gloucester products of the mineral fields of Shropshire, the manufactures of Staffordshire and produce of the farms of Worcestershire and Warwickshire. Shipping traffic had outgrown the port of Sharpness, its tidal basin crammed with vessels too long for the entrance locks of the Gloucester & Berkeley Canal. In the 36 days from 18th September till 24th October, 1868, so great was the overcrowding of the tidal basin at Sharpness that 136 ships for Gloucester sustained an aggregate detention in the Severn for 699 days waiting for admission, while some were kept waiting for 12 to 23 days!

W. B. Clegram, the canal's engineer, proposed a new entrance to the canal, from a quiet bay half a miles south of the original entrance, to provide a new entrance 300 feet wide between the two piers and allowing three-quarters of an hour longer for getting vessels in or out at each high tide than could be done with the old entrance. Inside would be a tidal basin 550 feet by 300 feet and further in a floating harbour dock 2,000 feet by 350 feet of sufficient width to turn the new steamships, finally tapering to 200 feet before joining the Gloucester & Berkeley Canal a short distance from the latter's original entrance.

Messrs George Wythes who had won the contract for constructing the dock (which was to be completed by 1st May, 1873) placed plant on the ground for a start in the spring of 1871 to turn what had been a primrose and cowslip meadow into a dock. Borings showed that the site was rock with clay above. Brick and lime kilns were set up, together with mortar mills and steam cranes for discharging stone and iron from vessels. Clay dug from the dock was used for making ordinary bricks, but blue bricks for corners and critical portions of the work came from Staffordshire. The lime was made from pebbles from Aberthaw burnt and ground on the spot as required. The contractor laid a temporary railway worked by steam locomotives.

Workmen were accommodated in huts, many of turf covered with roofing felt. A chapel was erected, a residential medical officer appointed and shops sprang up nearby. In 1872 smallpox broke out so a hospital was constructed and sufferers isolated. Agricultural labourers who tried navvy work found that their strength usually proved inadequate to dig the tenacious clay and wheel a laden barrow as easily as a professional navvy. They generally returned to farm work.

On 21st March, 1874 a crack appeared in the wall on the north side of the entrance lock. It grew in extent, causing 40 feet of wall to collapse. Water to rushed in and ripped one of the iron lock gates from its hinges. While excavating the dock, there were several interesting finds: a submerged peat forest, deer antlers, a horse's skull, a dog's skull and a boar's tusk.

The dock was completed about 10th November, 1874. On the opening day, 28th November, 1874, as there was no suitably timed train from Gloucester to Berkeley Road station, the directors hired a special to leave Gloucester at 6.00 am. It covered the 15 miles in less than 30 minutes before the 200 visitors were transferred to horse-drawn carriages, (the railway had yet to be opened), to take them to Sharpness.

On a wet and stormy morning, the first vessel was expected to arrive at 8.00 am. Captain Calway, harbour master, checked that the tow ropes and capstans were in working order. The first ship to enter was the *Director* adorned with flags from stem to stern and carrying timber from New Brunswick drawn by the tug *Milo* assisted by another. By now a crowd of a thousand had gathered. Next was *Tre Fratelli* an Italian barque with grain, then *Vaza* followed by *Protector* carrying Norwegian timber, the latter hauled by the tugs *Cambria* and *Vanguard*. As she entered the lock gates, *Protector* gave a salute from her two guns while her crew cheered.

By an Act of 25th July, 1872 the MR was authorised to construct a branch line four miles long from its main line at Berkeley Road to the docks at Sharpness. Having taken nearly three years to construct at a rate of just over a mile a year, it opened to goods traffic on 2nd August, 1875. There was little ceremony, the MR being represented by Mr Hunt, the goods depot manager, Gloucester; Mr Glover of the Gloucester locomotive department; Inspector Allard from Derby and Mr Mousley the sub-contractor. The first train to Sharpness was welcomed by W. B. Clegram the dock company's engineer and Captain Calway, harbour master. On its return to Berkeley Road, Mousley entertained a select party 'to an impromptu but capital dinner' provided by Gregory of the Prince of Wales Hotel. It was noted that the alterations to Berkeley Road station, necessary to accommodate passengers, were proceeding, though in the event the branch to Sharpness did not open to passenger traffic until 1st August, 1876.

The Severn Bridge Railway (SBR) called its first meeting at the Canal Office, Gloucester, on Saturday 31st August, 1872, William Charles Lucy, a Gloucester corn merchant and chairman of the Gloucester & Berkeley Canal being appointed chairman of the Severn Bridge Railway. It was stated that although the Midland Railway (MR) and Great Western Railway (GWR) had been approached, only the former approved the venture, but it was anticipated that the GWR would give its approval to the scheme by the time the work was completed.

The estimated cost of the railway was £277,973, the MR allowed to subscribe £50,000 towards the total. The GWR challenged its power to do so, but arbitration decided in favour of the MR. The GWR did not use its privilege of subscribing, although at one time it would have contributed if it could have excluded the MR and secured the sole running powers, but the GWR was already committed to funding the Severn Tunnel. The Gloucester & Berkeley Canal Company subscribed £50,000 and the Severn & Wye Railway half that amount. The meeting decided that a new prospectus should be issued inviting the public to subscribe to the rest of the share capital. Lucy announced that the SBR bill had passed Parliament in the face of formidable opposition because it was seen that the work would affect public good and 'even supposing the Severn Tunnel was made, many passengers would prefer passing over the river to going under it' (applause).

Lucy warned shareholders that work would not start immediately as the labour market was 'in great difficulties and prices of materials in a very unsatisfactory state'; he believed it wiser to wait and then place contracts when the labour market was satisfactory.

George William Keeling and George Wells Owen were appointed engineers with Thomas Elliott Harrison, President of the Institution of Civil Engineers in 1873, as consulting engineer. Harrison had trained under Thomas Blackwell, engineer to the Kennet & Avon Canal and Bristol Docks, who became consultant to the Severn & Wye in 1847. In 1860 Keeling succeeded Blackwell as engineer of the Severn & Wye Railway and extended and converted it to become a passenger carrying railway in 1875, thus opening up the Forest of Dean to tourists. On the absorption of the Severn & Wye in 1894 by the GWR and MR jointly, he became engineer of the Gloucester and Hereford Division of the GWR. It was while in this post be met with a serious accident.

At 7.35 am on 5th February, 1903 he left Gloucester in the inspection saloon hauled by a '517' class 0-4-2T. After inspecting Lydney Docks he proceeded north, but as he had been delayed by heavy traffic decided to return via the Mineral Loop to Foxes Bridge Colliery. This loop was operated without normal signalling and trains could, and did, enter from both ends. Porter-signalman Ellaway at Drybrook Road overlooked the presence of a coal train headed by '2021' class 0-6-0ST No. 2032 and allowed the special to enter the section without warning its driver.

At low speed the two trains met on a curve near Crump Meadow. The footplatemen jumped off before the collision, but inside the inspection saloon officials were thrown from their chairs and knocked unconscious. Unfortunately Keeling was standing on the observation platform with his back to a large window and was hurled through the glass.

An engine and brake van were commandeered from a ballast train standing in the station loop and this conveyed Keeling and his surveyor to Lydney where a doctor joined the train before it ran through to Cheltenham. Keeling's skull had been fractured and he did not regain consciousness until next day.

The subsequent inquiry revealed that Drybrook Road was a busy station: in addition to the special, the coal train and the ballast train, another coal train was in Trafalgar siding and a Cinderford to Lydbrook passenger train was approaching. As he spent 10 hours daily in his box, the following year the busy porter-signalman there was regraded signalman.

Keeling retired from the railway on 30th September, 1904, his last task being to report on the condition of the Severn Bridge. He died age 74 on 21st June, 1913.

Thomas Elliott Harrison, born in 1808 at Fulham, had begun an apprenticeship in civil engineering with William Chapman. In 1830 he was employed by Robert Stephenson in preparing plans for the London & Birmingham Railway and then engaged in the construction of lines in the north-east. He took a leading part in uniting various companies to become the North Eastern Railway and in 1854 became its chief engineer; his particular expertise was bridges. He died in 1888.

The prospectus stated that the Severn Bridge Railway was to run from junctions with the GWR and Severn & Wye Railway at Lydney, thus connecting with nearly every colliery and iron works in the Forest of Dean, across the river to Sharpness linking it with both the MR and the Gloucester & Berkeley Canal; the line would be about five miles in length. The company had powers to raise £300,000: £225,000 in shares and £75,000 in loans, but in practice the bridge and its associated railway cost £410,000.

In the preceding five years, imports at Sharpness had exceeded exports by an annual average of over 231,000 tons, the excess being caused due to discharged vessels proceeding in ballast to South Wales, or a Forest port, to take on an export cargo, generally coal or iron.

The same year, an Act had also been passed for cutting the Severn Tunnel, but the distance from South Wales to London was practically the same by both routes and the bridge gradients favoured that course. Should both crossings be completed, the Bridge directors were of the opinion that the important mineral traffic to London could be carried via the Severn Bridge at as low, or even lower price than by the Severn Tunnel.

At the shareholders' meeting on 28th February, 1875 William Lucy chairman, announced that all shares had been taken, the public having bought those which the GWR had declined. The contracts for building the bridge and railway in three years were in a forward state, but not yet signed. In March 1875 the directors were able to announce that the contracts had been placed.

The first stone, two tons in weight, was laid ceremonially at Purton on 3rd June, 1875 'in a quiet and unostentatious manner'. No public demonstration was expected, but rumour of the stone laying had spread and a large number of inhabitants gathered on the Forest side at short distance from Awre station.

Lucy and the other directors arrived at Awre in the noon train from Gloucester and held a meeting in the engineers' room at Purton Manor, reputedly once the home of Sir Walter Raleigh. Lucy said that an estimated 600,000 tons of coal and iron would cross the bridge annually and at 6*d.* a ton, less 40 per cent working expenses, would give a return of 3 per cent on capital.

At 2.00 pm the block was laid on solid rock by Lucy who then stood on it and gave a short speech. Following three cheers for the Severn Bridge and another three for the directors, the party enjoyed lunch in the engineers' rooms at Purton Manor. The engineers had erected a footbridge over the South Wales line to offer easy access to the shore. Following the meal, a special train carried the directors to Gloucester via Lydbrook and Ross. That evening a large group assembled at the Purton Passage Inn, the landlord providing a brass band to liven the proceedings.

The bridge contract was won by Hamilton's Windsor Iron Works, Garston, Liverpool, for £190,000. This involved the erection and founding of the pier cylinders, erecting and riveting 21 spans, building a swing bridge over the canal and another on the North Dock Branch adjacent to the New Docks. Samuel Sharrock, a civil engineer and manager for the Iron Works, appointed George Earle, who had worked for the company for 10 years, as bridge manager due to his experience of erecting difficult structures.

Messrs Vickers & Cooke, London, were given the £90,000 contract for building the Sharpness North Docks branch viaduct, the masonry and piers for both swing bridges, the Sharpness abutments, Pier No. 1, Purton Viaduct and the 506-yard long Severn Bridge Tunnel together with under line bridges, cuttings and stations.

Chapter Two

Construction Begins

Prior to the half-yearly meeting on 31st August, 1875, Clegram, the canal's engineer, presented Lucy with a silver trowel as a memento from the company directors for laying the foundation stone, while the contractors Vickers & Cooke, gave him another for laying the first stone of the north abutment pier.

Keeling, the railway engineer, reported that nearly 100 pier cylinders, 4 feet in length, 1¼ – 1½ inches thick and 6 feet in diameter, had been cast, together with a similar number of cast-iron cross bracings to join the cylinders together. Some of the 9 and 10 feet diameter cylinders had also been cast. The castings and wrought-iron superstructure of the swing bridge was in hand and work had begun on the wrought-ironwork of the 134 feet spans.

Previous long bridges, such as those over the Wye at Chepstow and the Tamar at Saltash, had been constructed on pontoons and then floated into position. Due to the strong Severn tides which could rise 30 feet in 2½ hours, this method was impracticable. Instead the cylinders were cast, the spans fabricated and erected at Garston before being dismantled and shipped to Sharpness in numbered sections then later moved to the site for erection. To speed assembly at the final location, the parts were temporarily bolted together, these bolts later removed and replaced by rivets when a span was securely in position.

Workshops had been erected by the bridge site at Sharpness, scows and barges constructed and plant provided. Piles for the stagings of two of the piers had been driven and approximately 50 cylinders had been delivered and were to be sunk immediately. The piers of the 312 feet spans, (bridge span measurements vary in different accounts: the present author has used those of the district civil engineer, BR, Gloucester), consisted of four cylinders each

On 8th May, 1877 George Keeling stands inside a 10 feet diameter pier cylinder as used for Piers Nos. 19 – 21.

BR WR

Pile driving: some men appear to be in perilous positions. *BR WR*

10 feet in diameter placed from foundation to low water level, and then 7 feet diameter from low water level to the top of the pier. The remaining spans, five of 174 feet and 14 of 134 feet, had two cylinders each respectively 9 feet diameter and 7 feet.

The width of the Severn at that point was 3,538 feet and that of the bridge, including the viaduct and swing bridge, 4,161 feet. Its total height from the deepest foundations in the rock to the base of the platform was 150 feet. The height of the large span above the platform was 39 feet and the headway at high water 70 feet. The bridge weighed 7,000 tons with 3,528 tons of wrought-iron in the girders. The cast-iron cylinders forming the piers were made in 4-feet lengths and when in position were filled with 4,321 cubic yards of lime concrete. The piers varied between 6-7 feet in diameter above water and 9-10 feet below. There were four cylinders in each of the three piers carrying spans over the navigation channel; the others had just two cylinders.

In April the contractors Vickers & Cooke commenced building the viaduct across the dock lands at Sharpness, the swing bridge over the canal and the masonry abutments of the bridge on the west shore. On 2nd July, 1875 they had possession of all the land required for the approaches on the Lydney side, work had begun on the deep cutting near Purton and part of the fencing erected.

At the meeting on 12th February, 1876 the engineer reported that works of the bridge and railway 'were making fair progress'. Sinking the bridge cylinders had commenced on 15th August, 1875, the work starting on the Sharpness shore where the currents were feeble and gave the contractors the advantage of working on a partially-completed structure when they reached deep water. Construction began on the east bank north of the bridge across a high sand bank known as 'The Ridge' and as during the period of the neap tide it was only covered by a few feet of water, progress was expected to be easy. Unfortunately the sand, 30 feet deep, held a high percentage of clay and difficulty was experienced driving piles for the pier supports, it being likened to penetrating a block of rubber. Fortunately help was at hand.

James Brunlees had encountered the same problem when constructing a viaduct across the sands of Morecambe Bay, so his plan was copied. A water jet provided by a donkey engine scoured away the sand so only a little force was required to drive the pile home. Piles were sunk from a pontoon fitted with a pile engine. The piles were lashed to the block of the pile engine to act as a guide, with a further guide lower down on the pontoon, the pile hanging with a loose chain so that its weight assisted the sinking. When sufficient piles had been driven, they were firmly braced and preparations made for sinking the cylinders.

The faces of the flanges were faced in special lathe. A little wet red lead paint then rendered the joints between the cylinders water-tight, this method being superior to that depending on a sheet of India rubber or felt between the flanges. It was a little more costly in the first place, but the result well worth the extra outlay. The engineers also believed that rickety scaffolding was false economy.

A few lengths of cylinder were fixed together and lowered from the staging on which had been placed an apparatus for gripping them and keeping them vertical. As sand was excavated from the interior of the cylinder it sank into position aided by being weighted with 150 tons of ballast. The air compressing apparatus was now ready. The air locks were the same as those used on Bouch's Tay Bridge. Men working inside the cylinders were subjected to a pressure of 5-40 pounds per square inch.

When a cylinder was well-founded on rock, it was lined with felt to give a flexible medium and filled with concrete; then when it had reached its correct height, a large stone weighing 5-6 tons was placed on top. Time proved that the novel method of lining cylinders above the foundations to allow for expansion and contraction was unsuccessful, because by 1960 most of the cylinders had been cracked by frost. The erection of the spans kept pace with the sinking of the cylinders.

The trow *Victoria* of Chepstow unloading barrels of cement, 23rd August, 1876. Ballast weights are hung on the left-hand pier to overcome the buoyancy of the compressed air. Later *Victoria* became wrecked after striking the bridge. *BR WR*

When the pier cylinders had been lowered to the bedrock and built up above high water level, the piers had to be founded into the solid rock. To clear water from a cylinder, a pneumatic apparatus was used. This consisted of a large iron air-bell and steam-driven air compressor. The bell had three chambers projecting from its sides, each with double doors to create an air lock, an externally-operated winch raising sand and rock extracted from the foundations.

The air compressor was similar to an ordinary pumping engine but pumped air instead water. When a pier cylinder was fixed in position and an air-bell attached to the top, all connections were made air-tight and the engine set to work pumping air in. The amount of pressure depended on the circumstances: if the cylinder was in shallow water little pressure was necessary to keep it dry, but if sunk well out in the river and borings were in rock, the pressure needed varied between 15 and 30 pounds per square inch.

Severn Bridge construction: view from Sharpness across the Gloucester & Berkeley Canal, with the base of the swing bridge in the foreground on the far bank. *BR WR*

Base of the swing bridge, left; the partly-sunken supporting columns of a span right; compressor on right of platform. *BR WR*

When water was forced out of the cylinders the excavators descended and dug out the sand which was drawn up and taken from the air-bell through double doors. The peculiar sensation experienced in working in compressed air was described as being very similar to that experienced in a diving bell, but any inconvenience was counterbalanced by comparative immunity from injury in a fall. To counterbalance the buoyancy of the compressed air, sections of rail were suspended from the sides of the pier. When founding the deep water piers some 70 feet below high water at spring tides, the pressure was raised considerably and the ballast weight increased to over 100 tons – more than was necessary but as a precaution against accident to life and property, the contractor being most careful and experiencing very few accidents, only one being fatal.

Following the extraction of the sand, the rock bed of the river was excavated to a depth of 4 feet using Reeves' pneumatic excavators, an awkward task in the confined space within the cylinders of only 9 feet and 10 feet in diameter. Workmen emerged from the air lock exhausted and bleeding from their ears and noses. As the excavation proceeded, the pier was inched down on the excavation, filled with concrete and the air pressure maintained and the air-bell not removed until the concrete had set.

When a stage had been erected, trusses were placed across reaching over a cylinder, thus forming a strong temporary platform. When that was done, the erection of a span was rapid work and on several occasions a small span was fitted into position for riveting within seven days. All riveting was done by hand and each rivet carefully tested.

The scaffolding needed for the erection of the first two spans was almost ready and seven spans 134 feet long had been delivered to the site at Sharpness and were ready for erection.

Two of the 174 feet spans were erected at Garston and the remainder of the small spans were in progress. The weight of the cast-iron delivered was 780 tons and wrought-iron a similar figure. The swing bridge across the canal had arrived. Part of the viaduct at Sharpness across the dock lands had been erected and the fencing almost completed throughout.

The longitudinal members of the staging were fitted with rails for a travelling crane for hoisting the bridge components into position. The bridge superstructure was simple: the top and bottom members of the bow string girders were connected by verticals which carried cross girders for the roadway. Small longitudinal girders between the cross girders were fixed to support the rails and the floor was filled with iron decking.

A steam-driven compressor fed the air-bell and piers. Sand and then excavated rock were removed by winch. Compressed air was maintained until the four feet of concrete foundations were set. *BR WR*

These various parts were initially bolted together as a temporary measure prior to riveting. Hand-operated forges heated the rivets which were set by blacksmiths and inspected after setting. The bridge was built on an incline which terminated at a summit in the centre of the Severn Bridge Tunnel.

The half-yearly meeting on 12th February, 1876 discussed whether the swing bridge should be constructed for double or single track. The chairman said the bridge approach was double and he believed it unwise to make the swing bridge single track as if doubling was required in the

Two spans on framework supporting the unfinished columns. BR WR

Four almost-completed spans. BR WR

View on 1st August, 1876 of the contractor's yard at Sharpness beside the Gloucester & Berkeley Canal. On the lower left are sections of tubes having arrived in railway wagons. *BR WR*

future, the stoppage of traffic on the swing bridge would be fatal to traffic. The estimated cost of the swing bridge was £15,000 for single track or £25,000 for double. Voting favoured that the swing bridge should be able to carry double track.

At the half-yearly meeting held on 5th August, 1876 the engineers reported that Hamilton Windsor's progress was satisfactory; sixteen piers were in the course of erection: eight founded on rock of which four were filled with concrete and been completed to the underside of the girders; seven piers had been sunk through sand to rock; Pier No. 21 was being sunk and the remaining five piers had not yet been commenced. Two spans had been erected and were complete. Regarding the railway approaches being constructed by Vickers & Cooke, progress was slow, so the SBR insisted that a locomotive be purchased to speed the work. To fulfil this demand, Vickers & Cooke purchased a new 0-6-0ST built by I. W. Boulton of Ashton-under-Lyne in 1876, utilising parts from one of the Sturrock Great Northern Railway steam tenders purchased in 1870. Named *Raven*, at the end of the contract it was purchased by the SBR in 1880 and sold in 1892. Vickers & Cooke's work rate was still unsatisfactory and in September 1876 only the heading of the tunnel had been made and work not started on the viaduct over the South Wales line.

The meeting resolved to borrow a sum of not exceeding £75,000 at a rate of interest not exceeding 5 per cent.

The *Gloucester Journal* of 26th August, 1876 reported:

ACCIDENT AT THE SEVERN BRIDGE

Some rather sensational paragraphs have appeared in some of the London newspapers this week about an accident at the Severn Bridge. The plain facts, however, are very simple. One of the piers – known as No. 15 – is being sunk in deep water, and one of its two cylinders was being founded by means of the pneumatic apparatus. It was known that the wood staging round the pier was not in good condition, owing to the wear and tear of the river, but it was hoped that it would last long enough to enable the other cylinder to be founded. But last Monday's tide [21 August] was exceedingly strong and the wind was rough and during the day the staging gave way and was carried some little distance, and the cylinder itself, despite the great weight on its base, heeled considerably over, though it was not overturned. Fortunately no one was injured by the accident, and though there will be some delay in the completion of the work, the loss is of labour only and not of material. It has been stated that the damage will amount to something like £2,000 but it will probably not exceed a quarter of that sum.

On 7th February, 1877 a special meeting considered whether a Bill be placed before Parliament seeking powers to raise additional capital for widening the swing bridge, providing additional protection to the piers and for the provision of stations, rails and signals. Additional time was sought to complete the project. An important clause in the Bill provided a guarantee of interest on borrowed capital to an amount not exceeding £75,000 jointly by the Midland Railway, Sharpness New Docks and the Gloucester & Birmingham Navigation which would allow the Bridge Company to borrow at a lower rate of interest than if using just the security of the Bridge Company. This Act was passed 2nd August, 1877, *40 & 41 Vic cap 148*. It also allowed extra time up to July 1880 for completion of the works which had been caused in consequence of the GWR not availing itself of the option to subscribe additional capital of £133,300 required for stations, rails, doubling of the swing span and other additional works.

At the half-yearly meeting on 21st February, 1877 the engineers reported that despite the unfavourable weather, satisfactory progress had been made, the cylinders for twelve piers founded and filled with concrete, three other piers were in progress and ten built up to full height and complete to the underside of the spans.

The girders resting on the cylinders were constructed of wrought-iron on a modification of the bowstring principle. Five spans had been erected and completely riveted and a further five in various stages of erection while the circular masonry pier to carry the swing bridge was well advanced. Cast-iron delivered was 2,800 tons and wrought-iron 2,400 tons. Two piers of the masonry viaduct at Purton had been started and the heading of the Severn Bridge Tunnel driven throughout, but Vickers & Cooke were still not making satisfactory progress.

At the half-yearly meeting on 18th August, 1877 the engineers reported that cylinders for fifteen piers had been founded on rock and thirteen spans erected, that is about half the length of the bridge. The swing bridge just awaited turning equipment. Sixty yards of the tunnel had been finished and although work on nine piers of the viaduct was in progress, it was not progressing at sufficient speed. Recent high tides had caused £200 damage to a pier.

On reaching Piers 14 & 15 from the Sharpness shore, the contractor had to contend with the full strength of the river at high water. Some very strong scaffolding had been erected for sinking these cylinders, but on one occasion during a spring tide the scaffolding was swept away and the partly-sunk cylinders capsized and lay in the river, divers having to separate all the cylinder joints before they could be salvaged.

Each stage of the replaced scaffolding had a base of approximately 150 feet and consisted of three rows of piles each 15 inches square driven initially by a gun-powder pile driver invented by an American, but later replaced by one of a Sisson & White design. The staging was cross-braced from top to bottom and protected by

With the Gloucester & Berkeley Canal in the foreground, work on the bridge is well under way. *BR WR*

cut-water dolphins which doubled as a safety precaution against being struck by a vessel.

The immense strength of the tide was demonstrated when the staging was being dismantled and having removed the cross-bracings, work was abandoned due to the turn of the tide. When work was resumed it was found that all the piles had been snapped off leaving just the stumps. On a further occasion two stones weighing about 6 tons, intended for placing on top of the piers, were left overnight on staging 8 feet above the low water line. Next morning the stones had disappeared and were found 60 feet away having been washed there by a high tide during the night.

On 8th September, 1877 Keeling reported 'a light locomotive belonging to our Company but not present in use' had been let to Vickers & Cooke at £14 a week.

It later transpired that Vickers & Cooke had secretly transferred their contract to John Brown of Bray. By February 1878 Brown asked to be released from the contract, so consequently the SBR took over his plant which was given to Griffith Griffiths of Lydney who was to complete the contract. Griffiths also built and fitted the three passenger stations.

At the half-yearly meeting on 16th February, 1878 the engineer reported that a total of nineteen piers had been founded and ten built up to the underside of the girders. Staging for the large pier had been constructed in deep water and four cylinders placed in position, the bridges and earthworks for the three miles from Lydney Junction were mostly completed.

Following the half-yearly meeting on 16th February, 1878, an extraordinary meeting opened to confirm the raising of additional funds authorised by the 1877 Act. These were £100,000 in £10 preference shares issued at £9 each as no initial interest would be paid until the completion of works. When £50,000 had been raised borrowing powers of £33,000 could be brought into operation.

On 17th February, 1878 a minor setback occurred when a strong tide resulted in the loss of intermediate staging consisting of wooden framing with strong iron braces erected between pillars No. 17 and No. 18.

With the Gloucester & Berkeley Canal in the foreground, the bridge is in an advanced state of construction. *BR WR*

View upstream *circa* 1876. Work on the bridge approaching the west bank of the Severn. *BR WR*

On 3rd August, 1878 the engineer reported that the difficulties experienced in founding the deep water piers had been successfully overcome and eighteen spans had been erected leaving only one 134 feet span and two of 312 feet to be put in place; the ironwork for the swing bridge across the Gloucester & Sharpness Canal was complete and the engines, boilers and machinery for turning it were being installed.

Griffith Griffiths of Lydney, who had taken over work on the approaches, formerly in the contract of Vickers & Cooke whose progress had been very sluggardly, had made good progress, two miles completed on the Lydney side and the cuttings at both ends of the tunnel almost finished. Only 50 yards of the 506 still needed to be excavated and lined and it was anticipated that this would be completed by September. The largest amount of work remaining was the masonry of the Purton Viaduct and earthworks on the Sharpness side were also behindhand.

Griffith Griffiths purchased from Fletcher, Jennings a new 0-4-0WT Works No. 153 for use on the contract. Named *Severn Bridge*, it was sold to the Severn Bridge Railway in 1879 which renamed it *Wye*, the name *Severn Bridge* given to a new Vulcan Foundry 0-6-0T.

The erection of the last three piers proved difficult due to the greater depth of water, so a higher pressure of 40 pounds per square inch was necessary within the cylinders to keep them watertight. This incurred additional ballast weights being hung round the piers to counteract any buoyant effects and to stabilise those piers against the early rush of the tides, thus giving a greater margin of safety to those working 70 feet below high water. This high pressure proved fortunate for one workman who fell over 70 feet to the rock, the very high pressure cushioning his fall to such an extent that the only injury he received was a broken arm.

Three months of frost during the winter stopped work on the viaduct, but work had restarted by 15th February, 1879, the date of the half-yearly meeting. This was followed by a special meeting held to put forward a Bill for permission to amalgamate the Severn & Wye Railway and the Severn Bridge Company.

The erection of the first 312 feet span was completed in February 1879 and although in planning it was believed that its deflection would be three inches, it actually only deflected ⅞ inch.

Staging for the last span began in March 1879. High tides in early April 1879 carried away scaffolding, but towards the end of the month was being replaced. Scaffolding in deep water had to be very substantial. By May work on the final 312 feet span containing over 500 tons or iron, was falling behind schedule. When cylinders were being sunk in the navigable channel, one was scoured out causing the cylinder to fall over, while on another occasion, two completed girder spans had fallen over in April when their framework was swept away by the incoming tide. Eventually more than 60,000 cubic feet of timber had to be used to support these two longest spans. Although work was now behind schedule, all was not lost, modern technology could assist.

Due to the strong flow of water to be covered by the two main spans, erection of the staging could only be carried out during neap tides, which meant that work was restricted to just seven or eight days each month. Francis William Thomas Brain, one of the proprietors of the Trafalgar coal mine was consulting engineer to the Electric Blasting Apparatus Company, Cinderford, so he installed two powerful floodlights powered by a 'Gramme machine', enabling a night shift to be employed on the bridge works. It is believed that this system was later used for a floodlit football match in the Forest, but this was not an innovation as floodlit soccer had taken place on the Spa Ground at Gloucester in 1877. It is interesting to record that in 1882 Brain installed the first underground electrically-driven pumps at Trafalgar Colliery.

On 29th April, 1879 the Bill for the amalgamation of the Severn & Wye Railway & Canal Company and the Severn Bridge Railway Company came before the Select Committee of the House of Commons. Arguments for the amalgamation were that as both railways were small, an amalgamation would be desirable for economy. Owners of mineral rights in the Forest were in favour of amalgamation.

The GWR opposed the preamble on the grounds that it had not been shown that the public would have gained advantage from the amalgamation and that the application was premature as the line had not been completed. The Committee found the preamble proved, the Bill receiving Royal Assent 21st July, 1879, *42 & 43 Vic Cap 163*.

The two companies agreed on the following terms:

> For the first two years after opening the Severn Bridge to traffic the net receipts should be apportioned 70 per cent to the Severn & Wye and 30 per cent to the Severn Bridge Railway;
>
> For the the third year the apportionment would be 55 per cent and 45 per cent respectively and in subsequent years each company would take 5 per cent. The capital accounts were to be kept separate and until 1885 the directors remained either in the Severn & Wye section or the bridge section.
>
> Initially the two companies would be known as the Severn & Wye Section and the Bridge Section, but on the opening of the bridge the undertaking to be called the Severn & Wye and Severn Bridge Railway. The Midland received running powers over the Severn & Wye, while the latter received running powers over the Midland from Sharpness to Nailsworth and traffic facilities to Bath and Bristol.

By the end of June considerable progress had been made since February and parties could traverse almost the entire length of the bridge along the permanent

The bridge is almost complete except for the few spans on the west bank. *BR WR*

Apart from the 312 feet spans, most of the bridge is complete. *BR WR*

way from the viaduct on the Forest shore. The only incomplete part was the final 312 feet span, but on 28th June, 1879 all its principals were in position and ready for the riveters to start work on Monday 30th June, this work being anticipated to take about a month.

At the half-yearly meeting on 2nd August, 1879 shareholders were told that the amalgamation of the Severn & Wye and Severn Bridge Railway had been approved and that the works were expected to be ready for a Board of Trade inspection that September. The engineer reported that Hamilton's Windsor Iron Works had nearly completed its contract. Both Lydney and the Severn Bridge passenger stations were 'in a forward state' while the joint station at Sharpness was nearing completion. One coal tip was ready to work within two months. The junction with the Great Western and Severn & Wye was ready for inspection, signalling on the Lydney side almost finished, and much of the permanent way had been laid. Despite unfavourable weather, Griffith Griffiths had made good progress and coping was being fixed to the viaduct on the Forest side.

The bridge was opened to the public every Monday 11th August until 15th September. At several weekends the bridge was open to public inspection, permits being issued by Earle at Purton Manor and Keeling at Lydney, while on the weekend before the opening day over 1,000 walked over the bridge.

In August the contractors began laying the permanent way, that over the bridge was carried on longitudinal timbers and comprised of 65 lb/yard Rhymney steel, hollow crown, running rails with 48 lb/yard 'T'-section guard rails as stipulated by the Board of Trade for this type of bridge. The track remained on the bridge until 1932 when it was replaced with 85 lb/yard running rail and the same weight for the guard rail. Initially the colour scheme for the bridge was black from the base of the piers up to the high water mark, with the remainder chocolate, while the spans were cream. The bridge deck was coated with a damp-resisting composition. The total weight of iron in the bridge was almost 8,000 tons.

View of the bridge and viaduct from Severn Bridge station. Notice the large blocks of stone for the piers, and the plankwork support for an arch in the foreground *BR WR*

Erecting 312 feet span No. 20, December 1878. *Author's collection*

Circa August 1879 with the bridge almost completed, workmen pose on Wheel Rock with George Keeling, centre front.
Peter White collection

On 15th September, 1879 the 0-6-0Ts *Will Scarlet* and *Friar Tuck* coupled together crossed the bridge to test the structure, then took 20 loaded wagons across on 2nd October, the day prior to the official inspection.

On 3rd and 4th October, 1879 Colonel F. A. Rich, carried out the Board of Trade Inspection. Eight heavy MR goods locomotives, a saloon carriage and another coach were involved in tests consisting of rolling and static loads applied to each span, rolling being conducted at various speeds. All the span girders were inspected at least twice. The deflection of the 312 feet spans was 1½ inches and the others ¾ inch.

The first passenger train to travel over the line from Lydney to Sharpness was a special for the directors, engineer, contractor, representatives of the Worcester & Birmingham Canal Company and the Gloucester station master. On 3rd September, 1879 this party left Gloucester at 11.15 am in a 'luxurious saloon carriage' placed at their disposal by the GWR. On arrival at Lydney it was detached from the train and shunted to the Severn Bridge line where a Severn & Wye locomotive, the 0-6-0WT *Forester* driven by William Ridler of Lydney, the company's chief driver, drew it to Sharpness where the canal directors proceeded to a meeting in their chairman's office while the remainder of the group explored the New Docks. *Forester* was a particularly unusual engine having been used on no less than three gauges: first on the 3 feet 8 inch gauge tramway with flangeless wheels; secondly on the 7 foot ¼ inch broad gauge and then finally on standard gauge.

At 2.30 pm all gathered on the Canal Company's steamer *Sabrina* for luncheon and then walked to Sharpness station to rejoin the special train. It moved onto the bridge and stopped allowing passengers to alight to inspect the work. After this the party divided: some returning to Gloucester on *Sabrina* while the others travelled by train via Lydney.

It was remarkable and a great credit to those in charge, that such a large undertaking was carried out at the cost of only three lives.

The first fatality occurred on 8th January, 1878, not on the bridge, but in the Severn Bridge Tunnel. John Tomkins was killed when the rope, securing a section of timber weighing approximately a ton, broke and it fell on him. On 1st March, 1879 William Aston was killed when operating a travelling crane in the construction of Purton Viaduct.

The crane had started from the centre of the viaduct and was carrying a very large stone towards the bridge when a strong wind blew for about 10 minutes, caught the crane and propelled it to the end of the viaduct. There it crashed into staging, was derailed and fell 70 feet to the ground. Aston tumbled 25 feet to the staging where he was seriously injured. He was moved to Severn Bridge station where he died early that afternoon.

Although the crane was spragged when stationary, it had no brakes to slow it down when in motion and the only method of stopping the machine was by the handles used for propulsion, adequate in normal circumstances, but no use in this instance.

On another occasion a workman fell from a pier and landed on sand. Presumed dead, as he was being taken away in a sack for burial he regained consciousness and lived to tell the tale.

The final tragedy occurred on 3rd June, 1879 when Thomas Roberts, engaged on the erection of the final 312 feet span, tumbled from rail level into the river. Falling, he struck the staging but was rescued in less than five minutes by a boatman who had witnessed his fall. He was taken to the engineers' room at Purton Manor and Mr Webb, a surgeon of Blakeney called, but Roberts died shortly after his arrival.

What may have been the first fatality to a sailor caused by the bridge occurred on 6th September, 1879. Thomas Shaw, of Gatcombe, a mile upstream from the bridge, had been to Avonmouth with his brother William and friend Thomas Margate to purchase an anchor. On reaching the bridge he planned to run through span No. 19.

There was little wind, but the rowing boat was caught by a huge eddy and turned broadside. Before evasive action could be taken it was swept into the timber staging and cut asunder. The three occupants clung to the staging but the timber supporting Thomas Shaw collapsed and he was swept away by the rising tide.

With great difficulty William Shaw and Thomas Margate climbed ladders and were able to reach track level, crossing the bridge to Sharpness where they were loaned a boat to recross the river to Gatcombe.

The bridge manager, George Earle, believing the body to be trapped in the staging, sent the company's divers down but were unsuccessful. The corpse was recovered on 9th September by David Long of Framilode higher up the river.

Chapter Three

The Bridge is Opened

The bridge, together with the 4 miles 7 chains Sharpness- Lydney Tin Works Junction with the Severn & Wye, the 10 chains Lydney Otters Pool Junction with the GWR and the 57 chains Sharpness Dock Branch, were opened to the public on Friday 17th October, 1879 with suitable ceremony for such an important structure. After a frosty night, although the day dawned bright, the prudent carried umbrellas and waterpoofs. These precautions proved needless.

The bridge company had invited about 350 significant guests – including the vice-consuls of France, Denmark, Germany, the Netherlands and the United States, all of whom were sent complimentary tickets by the Midland Railway to travel in a special train from Gloucester comprising 23 first-class carriages holding a total of 400 passengers. It was scheduled to leave at 11.00 am but was delayed due to the late arrival of the timetabled train to Bristol which it was booked to follow. A first-class coach and saloon were attached to the 10.15 am ex-Bristol Temple Meads and at Berkeley Road these were coupled to the special from Gloucester.

The special arrived at the east end of the bridge just before noon and was greeted by throngs of sightseers as it rolled onto the new and imposing structure. The *Gloucester Journal* poetically commented:

> The scene was very impressive. Below lay a vast expanse of yellow sand, [the tide was out], shimmering in the fitful rays of a sun which was determined to break through the curtain of cloud; around upon the green slopes congregated a multitudinous crowd of sight-seers and stretching to the farther shore was the solid yet graceful structure of the new bridge, strangely combining strength with beauty and presenting the maximum of stability with the minimum of obstruction to the waterway.

At Lydney the driver and fireman were joined on the footplate by the chairman, together with the Earl of Bathurst, the Lord Lieutenant the Earl of Ducie, the company's engineer and two Midland Railway officials making it very congested. As the train recrossed the bridge it exploded a fog detonator placed on each of the 21 spans thus giving a suitable royal salute. Apart from that over the Tay, it was the longest bridge in the United Kingdom as the Forth Bridge was yet to be opened.

Looking south to Sharpness, the opening of Severn Bridge from the *Illustrated London News*, October 1879. *Oakwood collection*

The Severn Bridge opened exactly 100 years after the very first iron bridge in the world was opened, also across the Severn, at Ironbridge, Shropshire.

After a brief stop at Sharpness the train crossed the bridge yet again where it stopped at Severn Bridge station. Most of the passengers alighted taking the opportunity to walk back to the first 312 feet span where the chairman, W. C. Lucy tightened the last bolt and in a stentorian voice declared the bridge open for traffic; for many years, this last bolt was painted red.

Following three cheers, many made their way on foot over to Sharpness station and onwards to the Pleasure Grounds where a large marquee had been erected and a magnificent luncheon prepared for the special guests by R. Fortt, Westgate Street, Gloucester; they began with hot soup – a splendid choice as it was a cold and windy day – and ended with tea and coffee. Among those present were Sir Daniel Gooch, chairman of the GWR.

An item which enlightened the rather boring speeches, was a toast Lucy made to the health of 'the non-subscribing company', meaning the GWR. In his speech the chairman told Sir Daniel that, should water enter the Seven Tunnel then being bored, he could offer the GWR a high and dry way over the Severn. Gooch replied politely and as luck would have it, while he was speaking, the GWR's route under the Severn received a severe set-back; navvies had released the Great Spring and flooded the tunnel workings.

Sir Daniel Gooch said that the GWR had never shown any antagonism towards the bridge and welcomed its construction, but circumstances prevented aid being given. He hoped the traffic of the bridge company would be developed with contributions from his company and other lines. With regard to the opening of the Severn Tunnel, (work on the tunnel had commenced about the same date as the bridge works started and Lucy had suggested that they might be invited to see the tunnel), Gooch, unaware of the flooding and anticipating that the two headings of the Severn Tunnel would meet in six weeks' time, cordially suggested that some of the present company might wish to accompany him in walking through it, but advised them to bring umbrellas! (Later in the proceedings the news that the Severn Tunnel headings had been flooded was given to Sir Daniel and this disaster delayed the tunnel project for a considerable period.)

Sir Daniel recorded in his diary:

> I went to the opening of the Severn Railway Bridge. There was a large gathering, and of course a feast. The day was cold and very uncomfortable.

A kind thought at the proceedings was a toast proposed to 'The Engineers and Contractors'.At the end of the speeches the party proceeded to Sharpness station and returned to Gloucester by special train.

It was not only the upper classes who enjoyed the day; Tom, Dick and Harry did too. The MR ran cheap day excursions from Gloucester to Sharpness down the left bank, while the GWR conveyed trippers down the other bank to Lydney. That afternoon visitors paid to walk over the bridge. Hundreds of Foresters swarmed to Purton and at Sharpness a fair was held in a field overlooking the docks. Ships in the docks displayed bunting and at each end, the bridge approaches were decorated with the flags of foreign nations.

Several trains of Forest coal crossed the bridge and corn sent from Sharpness to South Wales was able to travel direct rather than via Gloucester. Passengers complained that the fare from Coleford to Bristol was 'rather high' and the GWR route via the Portskewett-New Passage ferry cheaper.

On the opening day the company's stock consisted of eight locomotives, 26 wagons, four first-class and four third-class coaches.

It is pleasant to record that all the hard work put it by George Earle, manager of Hamilton's Windsor Iron Works Company, was not overlooked. On 15th October, 1879 the Bridge directors presented him with a large gold hunting watch supplied by S. C. Mann, Gloucester and inscribed:

> Presented to George Earle, by the Directors of the Severn Bridge Railway Company in recognition of his zeal and ability and uniform cheerfulness in carrying out the works of the Severn Bridge, under exceptional difficulties for the Hamilton's Windsor Iron Works Co.

W. G. Clegram, vice-chairman, paid glowing tribute to Earle saying that on numerous occasions he had seen him confronted with immense problems which at times were as much as he could bear, but the cheerful way in which he approached them and the way he handled his men in hazardous and difficult conditions, especially when founding and erecting the pier cylinders, earned him the respect of everyone connected with the project. Both Hamilton's Windsor Iron Works Company and the Severn Bridge Railway Company were most fortunate in obtaining the services of a man of his calibre.

On 17th October, 1879 Keeling gave a supper at the King's Head Inn, Blakeney to the 60-plus bridge employees whose services were terminated, while on 22nd October, 1879 Hamilton's bridge manager, George Earle, who had lived in Blakeney while carrying out the project, was invited to a public dinner and given another gold watch and chain. The Revd. A. D. Pringle had taken the kind and thoughtful initiative in organising a public subscription to cover its cost at the Bird-in-Hand Inn, Blakeney.

Shortly after the bridge's opening, Keeling himself was to be entertained by a banquet at Lydney, but owing to the death of the director Henry Crawshay this was deferred until 1st May, 1880. A testimonial paid for by the residents of Lydney and district, made by Messrs Martin & Son, Cheltenham and costing 100 guineas, was in the form of an elegant silver centre piece for candelabrum with glasses for flowers. Its base consisted of a plateau on one side of which was engraved the outline of the Severn Bridge, while the opposite side bore an inscription:

> Presented to George William Keeling, Esq., Civil Engineer, by his friends and neighbours, to record their hearty congratulations on his successful completion of the Severn Bridge, opened on 17th October, 1879.

At the ceremony Keeling revealed that all men connected with the river: pilots and masters among others, had entertained great prejudice against the bridge and as each pier was erected shouted on passing: 'When you come to the next one you'll find your mistake.'

The superseded ferry across the Severn from Sharpness had been worked by the Inman family for 200 years. The kindly SBR gave the final owner 'an appointment' so it was not necessary for him to join the ranks of the unemployed. He was given the responsibility of maintaining the navigation lights above the main channel.

The real beginning of the Severn Bridge was on Monday 20th October, 1879 when it was opened to regular traffic.

The main reason for building the Severn Bridge was to be able to easily transport coal across the Severn, yet while the bridge was being built, the Forest of Dean coalfield was experiencing trouble. Miners had refused to accept wages cuts in the early 1870s and either struck or worked part-time. When the Severn Bridge project was in progress this provided work relieving some distress in the district, but with its completion numerous families faced starvation.

Many acts of violence were perpetrated. One miner living near Blakeney who in an attempt to diversify kept a few sheep, awoke one morning to find all his animals had been brutally slaughtered and their legs broken. Similarly the landlord of the King's Head Inn, Blakeney discovered two of his cows' tails had been cut off in the night and had bled to death.

Relief schemes were far from satisfactory: under the Poor Law out of work miners were paid 1*s.* 6*d.* a day for constructing roads, yet their earnings as miners had been as much as 10*s.* a day. Many refused to accept these terms and some who did were abused by their friends. One kicked insensible caused Inspector Chipp of Lydney to investigate the crime.

Stonebreakers were employed at quarries to provide road stone, the rate for normal workers being 1*s.* 6*d.* a cubic yard, yet under the Poor Law they only received 9*d.* per cubic yard, but if a worker had a wife, an additional 2*d.* was paid for each child up to the age of 14. Much of this money was provided by charitable institutions such as churches, so when funds ran low, work had to be suspended until sufficient money had been accumulated.

In May 1879 216 sacks of potatoes were distributed to the starving at Parkend and 500 women arrived armed with every receptacle they could find to carry their allocation.

Following the Tay Bridge Disaster on 28th December, 1879, G. W. Keeling, the Severn Bridge's engineer, wrote a letter to *The Times* reassuring the public as to his bridge's safety. Lord Powerscourt replied, but unfortunately misinterpreted Keeling's facts, so to clarify matters Samuel Sharrock, manager of works and engineer of Hamilton's Windsor Iron Works Limited sent a letter dated 1st January, 1880 pointing out that the cylindrical piers of the Severn Bridge were virtually one solid structure. Every part of the bridge from the highest point to its solid rock foundation also had the lateral strength necessary to resist the violent winds of the Severn Estuary. The Severn Bridge had been examined by an independent engineer and found to be quite safe.

The first half-yearly meeting of the combined companies, now entitled the Severn & Wye and Severn Bridge Railway Company, was held at the Canal Office, Gloucester on 25th February, 1880. The chairman reported that the company had not been able to fully utilise Sharpness Docks for the expected coal trade owing to the coal tip at Sharpness not having been completed until recently. The North Dock coal staithe No. 1 was first used 8th January, 1880. A wagon turntable was essential as wagons were emptied through an opening hinged at only one end. This meant that the coal bunkering trade which was believed to be so important for the company had only just started. During December coal traffic over the bridge had decreased owing to a rise in its price. Lucy believed that it would take up to two years to develop trade over the bridge and that shareholders should be patient. Coal trade from the Forest had shown a considerable increase to the end of December, but was then checked by this price rise and the fact that one colliery was on strike and 'recently not a single ounce of traffic had gone over the bridge in the London direction'. The problem throughout the life of the bridge company was that matters depended very much on circumstances outside the directors' control.

Lucy reported that working expenses had been heavy, but net revenue had been sufficient to meet the fixed charges, leaving a small balance. The £33,000 loan capital, which had been authorised to be raised at the last meeting, was being rapidly taken up.

The engineer reported that work on the timber dolphins for guiding vessels past the channel piers and continuing the bracing between the pier columns to a greater height, was proceeding.

A portion of the main line had been relaid with steel rails replacing those of iron. Henry Bessemer had invented his converter in 1856 allowing steel to be manufactured economically and the Severn & Wye and

Severn Bridge Railways were taking advantage of having these better rails. Two new locomotives, *Sharpness* and *Severn Bridge*, ordered from the Vulcan Foundry were to be delivered in March 1880. More rolling stock had been ordered from the Bristol Wagon Company: passenger coaches, vans, horse boxes, cattle trucks, carriage trucks, wagons and brake vans.

On 29th June, 1880 Messrs Stephenson, Alexander & Co, auctioned the contractor's timber, materials and plant at Purton Passage and on 30th June the machinery and plant at Sharpness, including two 0-4-0WTs Fletcher, Jennings locomotives, No. 2 *Little John*, No. 3 and No. 5 0-6-0ST *Forester*. On 1st July the timber and remaining plant were auctioned at Sharpness. The three locomotives received no bids, but in July 1881 Keeling reported that No. 2 and No. 3 had been sold for £500.

At the half-yearly meeting on 25th August, 1880, Lucy, the chairman, told shareholders that he very much regretted that the accounts were not more favourable. He reminded them that at the last meeting he had predicted that it would be two years before traffic could be fully developed. Expenditure for the past half-year had been exceptionally heavy, partly from having to hire an additional locomotive. Arrangements had been made for developing traffic in steam coal from South Wales for loading into vessels at Sharpness Docks thereby saving 6*d.* a ton compared with the cost of leaving Sharpness in ballast to coal at Newport. The high-level coal tips at Sharpness were working efficiently and some cargoes of Welsh steam coal had been exported to colonial and foreign ports. Although trade in Forest coal was depressed,

MIDLAND
AND
SEVERN, WYE, AND SEVERN BRIDGE
RAILWAYS.
PLEASURE PARTIES,
FROM 1ST JUNE TO 31ST OCTOBER, 1880.
CHEAP RETURN TICKETS,
At 1¼ Single Fares,
WILL be issued from Midland Stations to PARTIES, not less than 6 First Class or 10 Third Class Passengers, desirous of taking PLEASURE EXCURSIONS to SEVERN BRIDGE, SPEECH HOUSE, or LYDBROOK.
The RETURN FARE for PLEASURE PARTIES from SHARPNESS to SEVERN BRIDGE will be—
1st Class, 1s.; 3rd Class, 6d.
By order. 3641

MR advert in the *Gloucester Journal* for an excursion to the Forest of Dean in the summer of 1880.

The barque *Wolfe*, built 1881, alongside the coal tip, Sharpness. This wooden structure was replaced by one of concrete in 1908.

Author's collection

there were indications of a revival. Despite the fall in coal tonnage, there had been a satisfying increase in other branches of goods traffic.

In November 1879 33,585 tons of coal had been carried, rising to 34,367 in December, but unfortunately due to this stimulation of the coal trade, the Forest of Dean mine owners raised the price with the consequence that a large proportion of coal from other districts filled the market. Coal tonnage carried by the railway was 24,191 for January 1880, 19,228 February, rising to 23,694 in March, 24,532 in April, 26,318 in May, and 32,793 in June.

Lucy went on to say that another reason why the company was not in so favourable position as expected was the small shipment of coal from Sharpness. It had been anticipated that the coalfields of South Wales would open a large trade to Sharpness, but a variety of circumstances had checked that development.

First, there was in America a 'ring' which had retained a large quantity of wheat and due to this, stocks of grain in the granaries of Gloucestershire were smaller than they had been for 20 years. He was glad to report that the 'ring' was now broken and that the import trade was now normal and that therefore there was a larger amount of shipping at Sharpness than since the docks were opened.

He hoped that captains might be induced to bring the coal freights back again instead of giving them to other ports. The railway had a very satisfactory arrangement with the GWR with reference to rates. It was able to ship coal at Sharpness at a saving of 6*d.* a ton to the ship owner as compared with taking on a cargo at Newport. In other words, the cost of sending a vessel in ballast from Sharpness to Newport was about 2*s.* 6*d.* ton and the difference in the cost of coal between purchase at Sharpness or Newport was only 2*s.* 0*d.* per ton, so that a captain would be 6*d.* per ton better off if he coaled at Sharpness rather than taking his vessel on to Newport.

Lucy went on to say that passenger traffic had shown a favourable improvement: in January 1880 receipts were £212 10*s.* 3*d.* but following this, the average per month was £242 11*s.* 4*d.* against £132 14*s.* 2*d.* for the corresponding previous half-year.

He said that the company had been short of locomotive power, so for 90 days an engine had been hired from the Midland Railway at cost of £2 per day. The company had now bought a new locomotive for £1,800 which would result in a consequent saving.

Lucy said that it was impossible to accurately estimate the future as the company depended on many circumstances over which it had no control, but he mentioned that the Midland Railway had an Act before Parliament for building a branch to Stroud. He anticipated a junction there with the GWR which would give the Bridge company access to the GWR on that bank of the Severn. His anticipations were not fulfilled: there was no junction at Stroud with the GWR, and the MR's Nailsworth and Stroud branches had little effect on the Severn Bridge Railway.

Keeling, engineer and general manager, reported to the meeting that the bridge was in good order and the bracing, between the columns and the dolphin fenders, mentioned in his last report, had been completed. The high-level coal tips at Sharpness were working efficiently and some cargoes of Welsh steam coal had been sent to colonial and foreign ports. The usual dredging had been executed in the canal and Lydney Docks. A further quantity of steel rails had been laid on the main line and some stations required a further siding and other accommodation. An iron shed measuring 132 feet by 70 feet for storing carriages and other rolling stock, had been erected at Lydney Junction on the site of the original Severn & Wye terminus.

The Great Blizzard of 1881 caused the Sharpness branch to be closed for two days and cost the company £150 for clearing the line of snow. Traffic was impeded for two weeks owing to blocks on other railways.

At the half-yearly meeting held on 25th February, 1881, the chairman was able to report that all the facilities were now available at Sharpness Docks, the dock company had reduced dock dues and that large shipments of Welsh coal were expected at Sharpness. The accounts to 31st December, 1880 showed a progressive increase in receipts sufficient to cover all the fixed charges and to leave a small balance. The depression in the iron trade was continuing and some works were closed.

Income from minerals and merchandise in the half-year to 30th June, 1880 had brought in £13,493 16*s.* 0*d.*, but this last half-year had earned £16,054 18*s.* 1*d.*, an increase of £2561 2*s.* 1*d.* Passenger and parcel receipts had been £1,726 19*s.* 3*d.*, but in the last half-year had risen to £2319 16*s.* 6*d.*, an increase of £592 17*s.* 3*d.* Total receipts, (including others than mentioned above), for the previous half-year had been £16,331 19*s.* 6*d.* and those for the present £19,546 19*s.* 4*d.*, an increase of £3,214 19*s.* 10*d.*

Expenditure for the 1880 half-year was £10,701 2*s.* 4*d.* and for the last £12,444 13*s.* 5*d.*, an increase of £1,743 11*s.* 1*d.*

The net receipts for the first half-year of 1880 after paying all expenses was £5,630 17*s.* 7*d.*, while the second half showed £7,102 5*s.* 11*d.*, an increase of £1,471 8*s.* 9*d.*

On Sunday 22nd May, 1881 the Portskewett Pier used for the GWR's cross-channel service caught fire, so a replacement service of four trains daily was run over the Severn Bridge, GWR engines working through and reversing at Berkeley Road. The service via the bridge which began on 25th May, was withdrawn on 15th June, 1881 when the pier was returned to use, as the GWR could not violate the steamer contract.

At the half-yearly meeting held on 26th August, 1881 the chairman reported an increase of £3,390 in receipts compared with the corresponding period for 1880, but with an increased expenditure of £2,218 4*s*. 1*d*. The tonnage of Welsh coal being shipped was improving: the half-year to June 1880 showed an increase of 1,012 tons; 1,820 tons to December 1880; 3,444 tons to 30th June, 1881 and 3,614 that July and August. He observed that the shipment of coal from Sharpness was in competition with Avonmouth and Portishead which charged lower rates. The meeting was not happy, a number of Severn & Wye Section shareholders complaining to the directors that they were disillusioned at the outcome of the amalgamation of the two companies. The engineer reported that the permanent way now consisted of double-headed steel rails, heavy chairs and creosoted sleepers, all more costly than those previously used.

A publication which the chairman described as 'a very useful and entertaining holiday guide book' extolling the beauties of the Forest of Dean had been compiled, 5,000 sold and a further 2,000 ordered from the printers.

At 8.00 am on 5th January, 1882 *Trotter*, a Gloucester-registered ship laden with tiles from Bridgwater, struck a pier of the Severn Bridge and foundered. Its pilot, William Clements of Newnham, steered the sinking vessel towards the shore. Captain Brinkworth, realising that his ship would not reach the bank, climbed the rigging with his crew. James White, aged 21 of Weston-super-Mare unable to swim, took to the lifeboat with a view to reach land, but unfortunately drowned when it was drawn down by the larger vessel.

The men in the rigging raised the alarm, were seen from the shore by Captain Petheram and James Morse who set off in *Sea Gull*, a pleasure boat owned by Mr Philpotts of Blakeney. With some difficulty they were rescued and taken to Captain Petheram's home for medical attention. Clements, the pilot, had a broken leg and was moved to Gloucester Infirmary; Captain Brinkworth and the other two men had a narrow escape.

At the half-yearly meeting held on 24th February, 1882, as receipts had increased by £2,000 over the corresponding previous half-year, a dividend of ½ per cent was declared, making 1 per cent for the whole year. B. S. Stock, a Severn & Wye Section shareholder, observed that the arrangements between the two sections, 70 per cent to the Severn & Wye and 30 per cent to the Severn Bridge, had resulted in a profit of £875 to the Bridge Section and a loss of £50 to the Severn & Wye.

The chairman explained that fortunately for the proprietors there was a very small proportion of loan capital on the Severn Bridge, but the Severn & Wye had £175,000 worth of debentures and certain fixed engagements which had to be discharged every year.

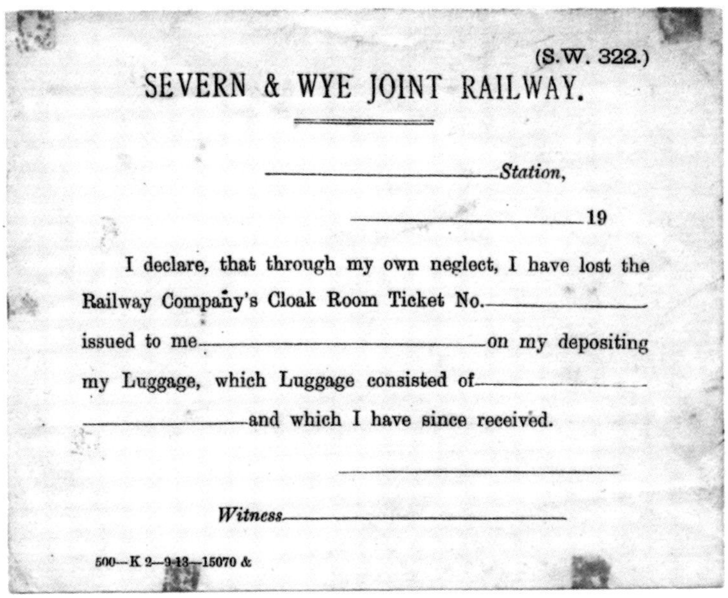

(S.W. 322.)

SEVERN & WYE JOINT RAILWAY.

__________ *Station*,

__________ 19

I declare, that through my own neglect, I have lost the Railway Company's Cloak Room Ticket No.__________ issued to me__________on my depositing my Luggage, which Luggage consisted of__________ __________and which I have since received.

____

*Witness*__________

500—K 2—9-13—15070 &

Severn & Wye Joint Railway passenger's acknowledgement of receipt of luggage.

Chapter Four

The Railway Seeks Expansion

The chairman at the meeting of 24th February, 1882 announced that two Bills were to be promoted in Parliament: the South Wales & Severn Bridge Railway and the Thames & Severn Railway, both of which would connect with their line. The first was to run from a junction 1½ miles from Lydney via Monmouth and Abergavenny to Talybont. The Thames & Severn Railway was to run from Stroud along the site of the Thames & Severn Canal and form a junction with the Swindon, Marlborough & Andover Railway to give access to Southampton and Portsmouth, while also offering a second route from South Wales to London via the London & South Western Railway. Alternatively it was possible that a line might be promoted from Stroud to the East Gloucestershire Railway, with running powers secured to Yarnton Junction, then having the option of either travelling over the London & North Western Railway to London, or going from Yarnton to Aylesbury and then reaching the capital via the Metropolitan Railway. These lines were promoted, unsuccessfully, in an attempt to provide a competitive route to London to break the GWR's monopoly. Had they been successful and therefore placed the Severn Bridge on an arterial through route, the company's fortune would have been made as it was through lines, rather than mere branches which created a profit. Part of the trouble was caused by South Wales colliery owners believing that railway companies should make railways rather than believing that they, too, should be involved.

On 16th March, 1882 the 0-6-0T *Robin Hood* travelling bunker-first on the 7.34 am Up passenger train from Lydney Town, emerged from Severn Bridge Tunnel in dense fog and struck the brake van of a train of Welsh coal which was fouling the loop points. Fortunately it was not derailed but a passenger was injured and damage cost the company £184 16*s.* 8*d.* At the ensuing inquiry, the Board of Trade officer Colonel Rich, censured the company for breaking regulations as the coal train driver carried no staff while the signalman and William Ridler, *Robin Hood's* driver were reprimanded for not being more observant.

It was the second time *Robin Hood* had been involved in a mishap, the first being in August 1881 when it derailed while drawing a passenger train. At the Board of Trade inquiry Colonel Rich concluded that the accident was due to poor permanent way, but added:

> It cannot be a very steady engine at any time when running fast and particularly so when attached to a light train, due to its short wheelbase, short springs and outside cylinders.

The next half-yearly meeting was held on 25th August, 1882 at the venue formerly favoured by the Severn & Wye Railway and Canal Company, the

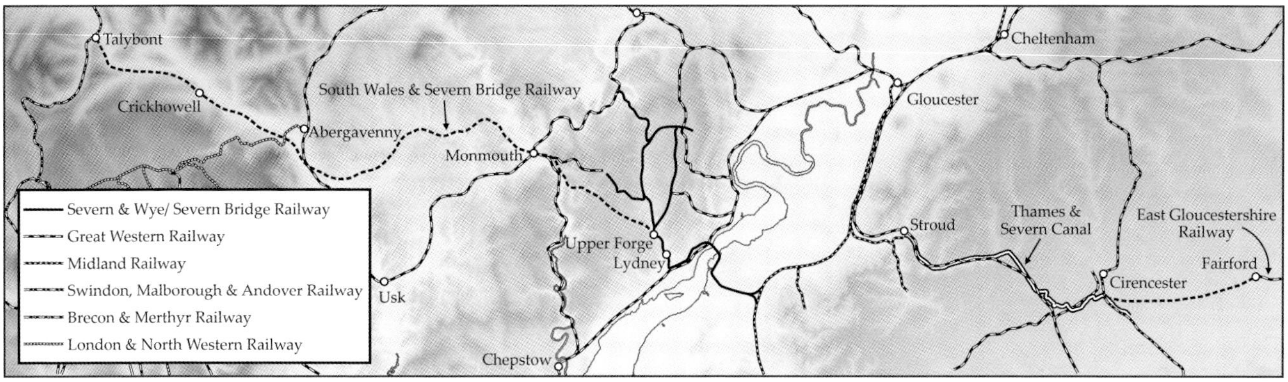

Sketch map of the proposed railways. In the west the South Wales & Severn Bridge Railway from Talybont to Upper Forge. To the east converting the Thames and Severn Canal from Stroud to Cirencester where it met the Swindon, Malborough & Andover Railway and the possiblity of connection to the East Gloucestershire Railway.
Relief shading contains OS data © Crown copyright and database right 2023.

Royal Hotel, College Green, Bristol; all subsequent meetings were held at this venue. The chairman, Lucy, expressed his great disappointment that plans mentioned at the previous meeting had been withdrawn for the two proposed lines which would have increased traffic over the bridge. He also announced that the last six months had shown a deficit of £1,463 0*s.* 8*d.* This was largely due to the exceptionally mild winter and the fact that large stocks of coal had been laid in the autumn resulting in trade in the early part of the year falling considerably. Receipts from passengers and parcels had remained about the same level as the previous corresponding period, while excursion and tourist traffic to the Forest of Dean had improved. In 1881 following the fire at Portskewett Pier, they had had the benefit of GWR passenger trains using the bridge for three weeks, but even leaving this out of consideration, passenger receipts had increased. £152 had been spent on a new dredger for Lydney Docks.

One shareholder queried the salaries of £600 and £400 per annum paid to the general manager and secretary respectively and also Keeling's pension of £533, while another observed that North country coal was sold at Bristol cheaper than that from the Forest of Dean. The chairman said that the object of the bridge was not to develop the Forest coalfield, but to advance Sharpness and the docks.

A shareholder, F. Roper moved 'That taking as an index of the incompetency of the directors of the Severn & Wye Section, the fact that no dividend had been paid for more than eight and a half-years to the original shareholders, a meeting be called for the purpose of appointing new directors for the Severn & Wye Section, or to adopt such measures as may be desired fit.'

Lucy said he thought it impossible to place this resolution before the meeting as the directors had been elected for a specific period and at the next meeting there would be an election of officers, the Severn & Wye directors then retiring and that would be the time for Roper to take action.

Stewart Fripp, another shareholder, said he was sorry that Roper had tried to bring charges against a body of men who having invested thousands, had done their best and devoted time an energy to revive the prosperity of the concern. He observed that Roper could have disposed of his shares many years ago at a good price, but had declined to do so. The meeting closed with vote of thanks from the chairman.

On 21st December, 1882 the promoters of yet another scheme to involve the Severn Bridge met at Cardiff, one being Lucy, the Severn Bridge Railway's chairman. The West of England and South Wales Railway was to run from Swansea to London via the Severn Bridge. A new line would be built from the MR at Nailsworth to Cirencester where it would seek running powers over the Swindon & Cheltenham Railway and the Swindon, Marlborough & Andover Railway to Andover and then use the LSWR to Waterloo. From Lydney a new line would be built to Bassaleg thus giving the company access to Cardiff and the Taff Vale Railway. As the scheme lacked encouragement from South Wales, on 9th January, 1883 the project was postponed.

At the half-yearly meeting held on 23rd February, 1883 the chairman reported that the accounts were the best since amalgamation, revenue amounting to £1,237, the improvement chiefly due to the carriage of minerals and merchandise. Passenger numbers had shown a slight increase. Although the shipment of Welsh coal was smaller than hoped, there was some increase – £18,199 against £11,476. The coal required at Sharpness was South Wales steam coal. Lucy said that a great many steamers were now coming to the docks there and had superseded the small sailing vessels. The latter had returned with a load of coal, whereas the steamers just took on sufficient bunker coal.

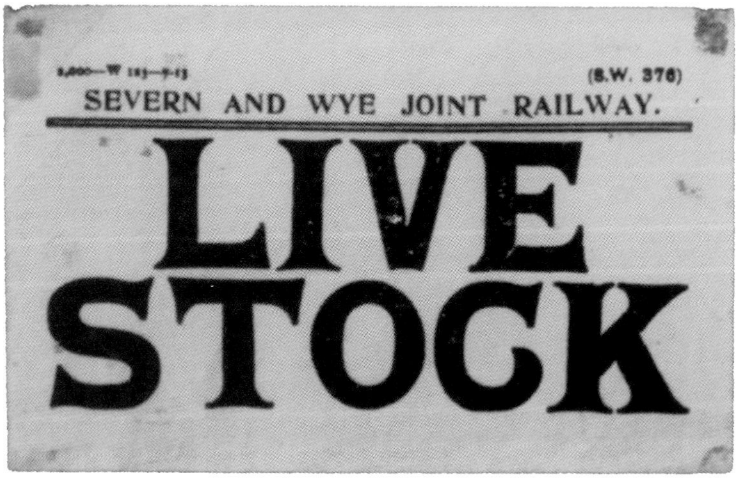

Severn & Wye Joint Railway goods label.

The bridge section of the line showed an increase of £837 4*s.* 7*d.* for the half-year, while the bridge had increased traffic on the Severn & Wye section by approximately 18½ per cent. Locomotive expenses had been reduced from 1*s.* 0*d.* per train/mile to 11½*d.*

The price of the carriage of coal to Bristol was revealing:

From Derby 6*s.* a ton
From Cardiff via the GWR 8*s.* a ton
From the Forest of Dean 4*s.* a ton.

The GWR had failed to divert its South Wales to Bristol coal trains from running via Gloucester, instead of using the shorter route over the Severn

Bridge, even though when the GWR's pier at Portskewett was destroyed by fire on 23rd May, 1881, thus preventing the cross-channel ferry service from being run, the Bridge Company had kindly provided an alternative route passenger service, (actually 25 minutes faster than the ferry), until temporary repairs permitted the pier to be used from 16th June, 1881. The GWR declined to give the Bridge Company powers to run two passenger trains from South Wales to Bristol.

G. W. Keeling reported that the maintenance expense had been approximately the same as the corresponding period in 1882, although it also included extensive repairs to coal-tips, bridges and so on and the task of relaying track with double-headed steel rails had continued. The cost of locomotive power had been 1¼*d.* per train/mile less than in the corresponding period, while carriage and wagon repairs were lower. The canal and docks had been dredged and maintained in good order. Telegraphic communication had been extended over the company's loop line where it had been greatly needed. Additional siding accommodation had been provided, a new engine, 0-6-0T *Sabrina*, purchased from the Vulcan Foundry and the bridge structure had been maintained in good order

2.000—W 135—6-14. (S.W. 311)

SEVERN & WYE JOINT RAILWAY.

No.

(Receipt number to be inserted here.)

LUGGAGE IN ADVANCE

From

To Station.

Excess Charge to pay

ONE OF THESE LABELS TO BE AFFIXED TO EACH PACKAGE.

Severn & Wye Joint Railway luggage label.

It was observed that the inland position advantage of Gloucester and Sharpness to a large extent was neutralised by the small difference in cartage rates from Avonmouth to Birmingham and Sharpness/ Gloucester to Birmingham, while the excellent warehouse accommodation at Sharpness and Gloucester was equalled or surpassed by that at Avonmouth. Small vessels which hitherto had sailed to Gloucester had almost disappeared from the corn and timber trade and the larger vessels did not proceed beyond Sharpness.

Grain could be discharged into a warehouse at Avonmouth for 1½*d.* less per quarter (28 pounds) than into a warehouse at Sharpness. At Avonmouth and Portishead unloading cargo and warehousing grain were both in the hands of the port authority, whereas at Sharpness and Gloucester it was carried out by private individuals and warehousing companies which proved more expensive than dock companies.

At the half-yearly meeting held on 31st August, 1883 the chairman announced with great regret that owing to the long continued strikes of the Forest colliers, the first quarter's receipts had suffered so severely that the directors were unable to renew or replace the debentures falling due in that half-year and required a payment of £73,000. The only answer was to place the company in Chancery and appoint the general manager and secretary joint receivers. The line's management would continue much as before, except that all receipts would be accounted for to the court which would periodically order a distribution of surplus funds. The amount of debentures falling due during that year was £73,000; of this £5,700 formed part of the bridge debentures and there was no difficulty in renewing these as they were guaranteed. In an effort to boost traffic, a reduction of 6*d.* per ton had been made on the coal rates from South Wales to Sharpness Docks.

Keeling reported that to economise the maintenance of the docks, canal and permanent way had been drastically reduced, but it had been necessary to take advantage of the fine weather to paint a considerable portion of the Severn Bridge. Sixty-five per cent of the main line and branches had so far been relaid with steel rails.

Lucy said that as a result of practising strict economy, expenditure during the last six months had been £1,017 less than the corresponding period the previous year, but receipts had fallen by £1,553. The Welsh coal traffic had not increased as expected but fallen to 6,862 tons from 7,478 tons. Arrangements had been made with the GWR whereby a reduction would be made in the rates and he hoped that this would increase the tonnage carried.

The coal tip was restricting the tonnage carried because it had been designed for filling smaller vessels, but now larger ships were using Sharpness and in the 11 weeks prior to 18th August, 1883, half of the vessels leaving Sharpness had too great a draught to load at the present tip. Should this trend continue, the directors would contact the canal company which owned the docks and request that a new tip be erected.

The chairman continued by saying that the company had purchased three locomotives – *Severn*

Bridge, *Sharpness* and *Sabrina* – and rolling stock for a total of £7,341 14*s*. 7*d*. and £1,707 8*s*. 3*d*. of this was still outstanding; sidings and new works had cost £14,000. The good news was that the July receipts were £3,635, while the average for the previous six months had been £2,854 and he anticipated that unless there was another colliery strike, income in the current half-year would cover the debentures.

At the half-yearly meeting held on 22nd February, 1884 the chairman reported that expenditure during the last half-year had been £13,438 2*s*. 9*d*. against £13,052 for the corresponding previous half-year. This increase had risen as a consequence of the considerable amount of dredging required at Lydney Basin and also having to replace iron with steel, rails.

After discharging and setting aside the amount required for the payment of debenture interest, there remained a balance of £932 to credit on the half-year's trading and if the whole year was taken, revenue was only short by £9.

To add to the company's woes, despite the reduction of cartage rates, the quantity of coal shipped at Sharpness had decreased. In 1883 the amount of tonnage leaving empty was 318,074 tons and of that tonnage, 207,750 tons were too deep or too long to proceed to the coal tip. Negotiations were to be made with the Gloucester & Berkeley Canal in the hope that some arrangement could be made for the provision of a further coal tip. The good news was that from 31st December, 1883 the GWR had carried South Wales steam coal to Southampton over the bridge, while one of the most productive tin works in the Forest was active again.

As the company was financially really two companies, a shareholder, H. Bennett proposed 'That in order to promote the success and future well-being of the entire undertaking, the directors be requested and are herein instructed to prepare and lay before the shareholders at a special meeting to be convened for the purpose at the Royal Hotel, College Green, Bristol, at as early a date as possible, a scheme for the reorganisation and amalgamation of all the stocks and shares in the undertaking, so that the two sections may be fused into one company with the assistance of Parliament or otherwise.'

Mr Stock seconded the motion and said that it righted a great wrong, it led to a divided board. They had begun last year with a debt owing from the Wye section to the Severn Bridge section of £2,200 which had now nearly reached £5,000.

A motion was carried 25:6 that the directors apply to the court for permission to form a fund to renew the company's locomotives and rolling stock.

The half-yearly meeting held on 22nd February, 1884 revealed that receipts and expenditure had been approximately on the same level as that of the corresponding previous half-year. After the payment of debenture interest there was a balance of £932. Notwithstanding a reduction in rates, shipment of coal at Sharpness had decreased from 10,721 to 8,600 tons. This had arisen due to the increase in the size of vessels. The total tonnage leaving Sharpness empty in 1883 had been 318,074 and of this 207,750 tons were either too deep or too long to use the existing tip. Negotiations were to be made with the Gloucester & Berkeley Canal in the hope that a new tip could be provided. One gleam of light was that one of the most important tinplate works in the Forest was active again.

At the meeting on 21st August, 1884 Lucy reported that receipts had been £3,000 more than in the corresponding previous half-year when they had experienced a Forest strike. A further light on the horizon was that the Woods & Forests Department of the Crown had introduced a Bill into Parliament that session for the purpose of facilitating the opening of deep mines in the Forest which would exploit seams similar to those providing steam coal in South Wales. The shipment of Welsh coal at Sharpness was still 'of limited character' owing to the need of a new tip for loading in deep water. Nevertheless 800 tons more had been carried than in the corresponding period of 1883, but the 21,075 tons sent to Southampton was lower figure than expected.

Keeling, the company's engineer and general manager, reported that expenditure had been 67½ per cent of receipts, about the same proportion as the previous corresponding half-year in which period due to the Forest strikes all possible expenditure such as relaying and so on had been suspended. It had been necessary to rebuild the swing bridge at the Lower Docks and part of this had been executed during the past half-year. Track relaying with steel had continued with greater rapidity. A further portion of the Severn Bridge had been repainted and the structure was in good order. The cost of locomotive power had been 11*d*. per train-mile.

Interestingly, due to its publicity value, Bellow's guidebook *A Week's Holiday in the Forest of Dean* received a Severn & Wye Railway royalty from 1884 until 1903.

The meeting held at Bristol on 28th February, 1885 revealed the good news that receipts had risen by £1,317, partly due to the fact that the shipment of Welsh coal at Sharpness had increased and it was hoped that a deep water tip could be installed so that the trend could continue. The GWR was sending Welsh coal to Portsmouth as well as Southampton via the Severn Bridge. The engineer reported that 77 per cent of the company's line had been laid with steel rails. Locomotive power had cost 12.1*d*. per train mile and he also suggested immediate provision be made for renewing some of the rolling stock.

The case for the amalgamation not working came before Mr Justice Chitty of the Chancery Division on

20th June, 1885 as the Bridge Section did not pay and the Severn &Wye Section was earning eight per cent of the profits. Twenty-three counsel and solicitors met to settle the case. A scheme of arrangement gave secured creditors about £1,000 a year interest; all the debenture stocks were to be consolidated; a new debenture stock to be issued at 4 per cent in lieu of the old and to be applied in paying off existing mortgage debts; one year's interest to be capitalised; the amount of new debenture slightly to exceed the amount necessary for paying off the existing mortgage debt; all actions to be stopped; the guarantees given by the Midland Railway and Gloucester & Birmingham Navigation Company to be written off; the existing share capital to be converted into Ordinary and Preference shares and to be apportioned among the shareholders according to their rights. The receiver was discharged from 31st December, 1884.

For those unacquainted with finance, it will be helpful to explain that there are broadly three ways of investing in a company:

1) Taking debentures – a loan paying a fixed interest – so your capital is fairly safe.
2) Buying Preference shares which entitled the holder to a fixed dividend, its payment taking priority over that of ordinary shares. This is a fairly safe form of investment.
3) Buying Ordinary shares – the amount of dividend varies according to the profit made. It could be risky or highly profitable.

At the meeting on 21st August, 1885 it was announced that passenger receipts were £50 lower, but mineral traffic showed an increase of £900. the tonnage of coal sent from Sharpness had risen from 7,734 to 14,281 and the Gloucester & Berkeley Canal Company was erecting a deep water coal tip. Completion was expected by the end of 1885. Keeling reported that a lathe had been purchased, a large expenditure that would effect a saving on repairs as hitherto railway wheels had to be sent a considerable distance for turning.

Complete financial fusion between the two sections of the company had been achieved under a Scheme of Arrangement enrolled in the Chancery Division of the High Court of Justice 21st July, 1885.

At the meeting on 26th February, 1886 it was revealed that the GWR had declined to agree through rates, so the Bridge directors proceeded under the Regulation of Railways Act 1873, to appeal to the Railway Commissioners. The hearing lasted five days and the judgement was 'generally satisfactory'.

The company had petitioned on the grounds of contention that as in 1872 Parliament had sanctioned a bridge and a tunnel, the Severn & Wye was entitled to have access to places on the other side of the river to which the tunnel also had access. The Severn & Wye said it did not seek any advantages, but in the public interest, any Forest of Dean trader who required to send produce should have the opportunity of sending it by either route.

Lucy reported that in effect the GWR said: "We have made the Tunnel and expended a great deal of money upon it, and we are entitled to the traffic," while the SBR directors claimed: "It is quite true you have expended the money, but we were in Parliament the same year as you were, and our expenditure upon the Severn Bridge, we consider, relatively to our capital, is as large and very much larger than the expenditure you have made upon the Severn Tunnel with reference to the Great Western Railway's Company's capital." Lucy said, without wishing to be harsh on the GWR, that it was one of those cases where a large company wished to avail themselves of the great powers they possessed to prevent the due development of a smaller company. He hoped it would now be bygones between themselves and the GWR. The GWR had fought every point and employed some of the finest counsels. The Severn Tunnel was opened to goods and mineral traffic on 1st September, 1886 and to passengers 1st December, 1886.

To accommodate the increased amount of shipping, in April 1886 the long-awaited new deep-water quay with hydraulic coal tip was opened at the western corner of Sharpness Docks to handle goods such as steel imported from France for Lydney Tinplate Works, sugar for Cadbury's factory at Frampton, the latter also using Forest coal brought by rail to Sharpness and then, using the timber-built coal tip opened 8th January, 1880, placed in dumb barges to be taken six miles along the Gloucester & Berkeley Canal to Frampton.

On 20th August, 1886 Lucy announced at the half-yearly meeting that gross receipts had been £21,850 compared with £21,328, but that owing to falling trade at Sharpness and fewer vessels calling, exports of Welsh coal had decreased to 6,700 tons. Expenditure had increased due to law charges for expenses in connection with the Railway Commissioners, the total cost being £1,750 of which the Severn & Wye's share was £612 17*s.* 3*d.* The company had petitioned on the grounds of contention. Keeling announced that 4¼ miles of steel rail remained to be relaid at a cost of £1,900 – £2,000 per mile.

At the half-yearly meeting on 25th February, 1887 it was revealed that the opening of the GWR's Severn Tunnel to goods and minerals on 1st September, 1886, had caused the Bridge Company's mineral receipts to fall from £19,861 to £18,947 and the coal rate to Bristol had been reduced to 3*s.* 2*d.* a ton. Passenger receipts which had been £2,492 the previous corresponding half-year, had fallen by £152. The maintenance of way and works had cost £6,954 compared with £7,679, while the cost of locomotive power had been reduced from £3,769 to £3,179, at 11*d.* per train/mile. Legal expenses had been £1,230 compared with £706.

Keeling reported that only a length of three miles of main line remained to be relaid with steel track. Repainting the Severn Bridge over four years had been completed.

It was reported at the half-yearly meeting held on 19th August, 1887 that receipts had been almost the same as the corresponding period the previous year, while expenses had been reduced by £406. The carriage replacement fund opened in 1886 with £30 stood at £531. The wharf, sidings and weighbridge at Oldminster Junction had been completed. Work had started on repainting the Severn Bridge for the second time. To facilitate train working, Tyer's Train Tablet Apparatus had been adopted for the single line Lydney Town – Tuft's Junction replacing Staff and Ticket Working.

At the half-yearly meeting held on 26th February, 1888 the chairman reported a fall of £1,654 in receipts due to the poor Forest coal trade – not to be wondered at as mine owners had raised the price of coal by 1*s.* a ton whereas in South Wales the price had only risen by 6*d.* a ton. Expenditure had decreased by £2,023 due to economy in maintenance.

As the GWR had declined to agree to the through rates, the Bridge directors had again proceeded under the Regulation of Railways Act 1873 to appeal to the Railway Commissioners. The hearing lasted five days and the judgement deemed generally satisfactory. The Commissioners granted the rates asked for, but they were to run for only a year, but with the option to apply again after the Severn Tunnel had been opened. As the GWR's tunnel offered a shorter route to the south and west of England, the rates were expected to be lower than the bridge's present charges.

Keeling reported that a new engine had been ordered from the Vulcan Foundry, that the cost of locomotives was 13½*d.* per train/mile, that the new deep water coal tip would open in March (it was actually first used in late April 1888) and that the original tip on the canal would be altered to more efficiently handle Forest coal. Dredging had been carried out at Lydney Canal and Docks, but the dry weather meant that steam pumps had to be installed at the docks to maintain the water level, water being drawn from the river when required. Only 2½ miles of track still required relaying with steel rails and the Severn Bridge structure was in good order. He said that the adoption of Tyer's Electric Train Tablet Apparatus was desirable for the other busy sections of line.

It was the same story at the meeting on 24th August, 1888 when Lucy had to reveal that receipts were down by £1,838 due to the depression of trade in the Forest together with intense competition while coal was arriving in Bristol from other areas. Expenditure had been £1,696 less and passenger and parcels receipts remained constant. Some passenger coaches had been extensively repaired, the cost being defrayed from the reserve. Shareholder H. Bennett said that a great deal of Forest coal was despatched via the Severn Tunnel rather than over the bridge and stated with pathos 'that it seemed as if the heart had been taken out of the shareholders; they felt themselves in a state of utter destitution.' He reminded the meeting that when he had told them that opening the Severn Tunnel would affect the position of the company his remarks had been laughed at.

Keeling, the engineer, reported that there had been a recent increase in Welsh and Forest coal to Sharpness. Painting the Severn Bridge was in progress, one of the swing bridge boilers had been overhauled and extensive repairs were being carried out to the lock and harbour gates at Lydney. Some passenger carriages were being extensively renovated and ⅓ mile of track relaid with steel rail leaving just over 2 miles to complete. The cost of locomotive power had been 11½*d.* per train-mile.

On 22nd February, 1889 at the half-yearly meeting Lucy reported a revival of trade and an increase in passenger revenue of £426 and £359 for goods, making the company's position better than three years previously, Tyer's Train Tablet Apparatus had very

SEVERN AND WYE AND SEVERN BRIDGE RAILWAY.—FOREST OF DEAN.

FROM JUNE 11th to Sept. 30th EXCURSION TICKETS are issued on MONDAYS and FRIDAYS by the Midland Railway from CHELTENHAM, GLOUCESTER, STONEHOUSE, DURSLEY, and CAM, to SEVERN BRIDGE, LYDNEY, SPEECH HOUSE ROAD, COLEFORD, and LYDBROOK JUNCTION. For times and fares see Midland Company's bills. Cheap Return Tickets will be issued to Pleasure Parties on any day for not less than Six First or Ten Third-Class Passengers. Special arrangements can be made for Schools, Benefit Societies, Working Men's Clubs, or Manufacturers' Trips, &c. The Speech House (lately considerably enlarged) is situated on a hill in the centre of the ancient Forest, and is a most attractive place for Pleasure Parties. Every accommodation can be provided at the Speech House for Picnics in the open Forest, and there are Cricket, Quoit, Archery, and Lawn Tennis Grounds. Telegraph Communication in the Hotel (address *via* Lydney). Lydbrook is beautifully situated on the River Wye amidst charming scenery. Boats can be provided on application by letter to Thomas Davies, Symonds Yat, Ross. Symonds Yat and Goodrich Castle are within a short distance from Lydbrook Junction Station. A Train at 7.0 p.m. from Speech House Road is run for Pleasure Parties when required, sufficient notice of which must be given. Tickets to Lydbrook for Symonds Yat are available to return from Coleford Station or *vice versa*, and Passengers can break the journey at Severn Bridge. An attractive Guide to the Forest of Dean is published by John Bellows, Gloucester, price 6d., and can be had at the Railway Stations.

MR advert in the *Gloucestershire Chronicle* for an excursion to the Forest of Dean in the summer of 1888.

much improved line working. Only approximately half a mile of steel track needed to be relaid and the work would be completed within a fortnight. The debt on *Forester* purchased in 1886 had almost been discharged and during the next half-year all the engines would be paid for and another ordered. It was hoped to use the reserve fund to pay a considerable proportion of its price. Four coaches had been renewed and two more were in the course of renovation.

On 23rd August, 1889 the good news continued as Sharpness Dock was reported to be full of vessels unloading grain and timber, while Bristol was still waiting for good times. Revenue had increased by £1,878 but unfortunately the Forest of Dean had raised the price of coal before the other districts. This seriously affected revenue as the SBR was a mineral line, minerals earning £19,516, passengers and mail only bringing in £2,104 – nine times less. Lucy said that if the Forest of Dean collieries reached the deep seams they would have access to steam coal which could be delivered to Sharpness and transported to other places at less cost than that sent from South Wales.

Lucy drew the meeting's attention to an article in *The Times*:

> The Grand Junction Railway proposes to run from the Metropolitan Railway at Aylesbury, through Oxford, Witney, Andoversford and across the Severn Bridge to the Forest of Dean, while the Golden Valley Railway was an ingenuous line which is good enough to offer five per cent debentures to shareholders of the Great Western and North Western on the most modest terms, has hopes that one day its fortunes may be more worthy of its title when it conveys from Hay through to Pontrilas on the route to Monmouth and the Severn Bridge all the wealth that the new Cambrian line can despatch to the metropolis.

On 28th February, 1890 Lucy announced that gross receipts had risen from £21,855 to £22,532 and of the latter, only £2,793 came from the carriage of passengers. An increase in the price of coal and iron had increased locomotive expenses. £500 had been added to the rolling stock fund which stood at £2,127 16*s.* 11*d.* The company now had 'a rather handsome saloon carriage' from the ex-Bristol Port Railway & Pier in place

Lydney Harbour crowded with ships. The wagons on the right belong to Parkend Deep Navigation Collieries. The cottage-style building to the left of the Parkend wagons was originally the Severn & Wye Railway offices.

Author's collection

237.—SEVERN AND WYE AND SEVERN BRIDGE.

An amalgamation of the "Severn and Wye" and "Severn Bridge" undertakings; the former incorporated as the "Lydney and Lidbrook" by act of 1809, and the latter by act of 18th July, 1872. The amalgamation was carried out under act of 21st July, 1879. The amounts received to June, 1889, on capital account, from share and debenture stocks, are as follow:—Share stocks, 623,084*l.*; debenture stocks, 321,191*l.*; total, 944,275*l.* The expenditure amounted to 941,053*l.*

The amalgamation took effect from the opening of the Severn Bridge Line, on the 17th October, 1879.—For other particulars relating to the past history of these companies, independently, see the MANUAL for 1879, and previous editions.

SCHEME OF ARRANGEMENT.—*Enrolled 21st July*, 1885.—Under this scheme the following new stocks were authorised, viz.:—

4 per cent. debenture stock (guaranteed by the Midland Railway and Sharpness New Docks, &c., companies)	£75,000
4 per cent. debenture stock	268,000
4 per cent. preference stock "A"	50,000
4 per cent. " " "B"	107,500
4 per cent. " " "C"	298,280
Ordinary stock	167,383
Total	£966,163

The above stocks absorbed the old shares and loans of the separate sections as follow:—All old mortgages, debentures, and debenture stocks received an equivalent of the new debenture stock at the following rates, viz.:—

5 per cent. loans to receive 110*l.* per cent. of new stock.
4¾ " " " 107½*l.* " "
4½ " " " 105*l.* " "
4 " " " par "

The new debenture stock bears interest at 4 per cent. per annum from 1st January, 1885. The arrears of interest for the year 1883 were capitalised as new debenture stock, and those for the year 1884 paid in cash.

The preference and ordinary capital consolidated and converted as follows:—

Old Shares.	Receive in Exchange.
Bridge Section 5 per cent. preference 10*l.* shares	5*l.* per share of 4 per cent. pref. stock "A"; 5*l.* 7*s.* 6*d.* " " " "B"
Wye Section 4½ per cent. guaranteed 20*l.* shares	22*l.* 10*s.* " " " "B"
Wye Section 5½ per cent. preference 10*l.* shares	10*l.* " " " "B"
Wye Section 5 per cent. preference 10*l.* shares	10*l.* " " " "C"
Bridge Section ordinary 10*l.* shares	6*l.* 10*s.* " " " "C"; 3*l.* 13*s.* 4*d.* per share of ordinary stock.
Wye Section ordinary 50*l.* shares	25*l.* per share of ordinary stock.

Voting powers to proprietors of preference stock "C" at the rate of 1 vote for every 100*l.* up to 5,000*l.* stock, and 1 vote for every 250*l.* stock beyond 5,000*l.* Proprietors of ordinary stock 1 vote for every 25*l.* stock up to 1,000*l.*, 1 vote for every 100*l.* stock above 1,000*l.* and up to 5,000*l.* stock, and 1 vote for every 250*l.* stock beyond 5,000*l.* stock. Proprietors of preference stocks "A" and "B" have no votes.

The full text of the scheme will be found on reference to the *Appendix* to the MANUAL for 1886, page 535; and for the constitution of the capital prior to the confirmation of the scheme, and other particulars, see the MANUAL for 1885, pages 293 and 294, and previous editions.

MILEAGE.—Authorised, 41½; constructed, 37; worked by engines, 33¾. Severn and Wye Original Line (Tramway) opened in 1813; new lines (locomotive) opened in 1872, 1874, and 1875; Severn Bridge Railway opened October 17th, 1879.

REVENUE.—The net earnings for the half-years ended 31st December, 1890, and 30th June, 1891, sufficed to pay the full dividend of 4 per cent. on the "A" preference stock, and ¾ per cent. on the "B" stock for the half-year ended 31st December, 1890.

TRANSFER DEPARTMENT.—Ordinary form of transfer; fee, 2*s*. 6*d*., certificates of all stocks required to accompany transfers; the several classes of stocks may be transferred on the same deed; no fractional portion of 1*l*. transferred.

Directors.—The board consists of 14 directors, one of whom is appointed by the Midland and one by the Sharpness Docks. *Qualification*, for the remaining 12 directors, 250*l*. of ordinary stock.

DIRECTORS:

Chairman—WILLIAM CHARLES LUCY, Esq., Brookthorpe, Gloucester.

Deputy-Chairman—JOHN A. GRAHAM-CLARKE, Esq., The Manor House, Frocester, near Stonehouse.

Steuart Fripp, Esq., Summerlands, Pembroke Road, Clifton.
George Baker Keeling, Esq., Severn House, Lydney.
Colonel E. A. Noel, Outwoods Hall, Duffield, Derby.
E. B. Merriman, Esq., 39, Threadneedle Street, E.C.
W. F. Brookman, Esq., Bristol.

Representing Midland Railway:
H. Tylston Hodgson, Esq., Harpenden, Herts.

T. Nelson Foster, Esq., Allt Dinas, Bayshill, Cheltenham.
Henry Wethered, Esq., Ferncliff, Tyndall's Park, Clifton.
Sir W. H. Marling, Bart., Stanley Park, Stroud.
O. A. Wyatt, Esq., Gibraltar, Monmouth.
Capt. W. B. Marling, Lydney.

Representing Sharpness Docks Company:
C. B. Walker, Esq., Gloucester.

OFFICERS.—Sec., Thomas Linton; Eng. and Gen. Man., G. W. Keeling; Auditors, Fred. A. Jenkins, F.C.A., Bristol, and W. Green, The Reddings, Cheltenham; Solicitors, Wintle and Son, Newnham.

Offices—Lydney, Gloucestershire.

Severn & Wye and Severn Bridge Railway Company details from *Bradshaw's Railway Manual, Shareholders' Guide and Directory*, 1892.

of the old directors' carriage which had required extensive repair. It could be hired and excursion parties could arrange for a man to be in charge of provisions and suchlike. The old saloon had become a brake van. Keeling reported that five more coaches needed restoration and some of this work would be carried out in the current half-year.

The Regulation of Railways Act, 1889 required certain appliances to perfect the block system, that is the need to interlock points and signals and that passenger trains should be fitted with continuous brakes. This Act authorised the company to raise capital for this expense by issuing debenture stock.

At the half-yearly meeting on 22nd August, 1890 Lucy was able to offer a more encouraging report. Gross receipts had shown an increase of £806, £200 of this coming from passenger traffic. A dividend of 2 per cent would be paid on Preference A Stock. He had hoped that iron ore traffic from the Forest would be developed, but this was now unlikely due to the introduction of the cheaper Spanish ore and the use of the Bessemer steel process. Until that date Forest of Dean ore having plastic qualities, had proved very valuable in mixing with other ores, but now that Bessemer steel could be formed into many shapes there was less call for Forest ore.

He observed that with only 37 miles of line, working expenses were heavier in relation to receipts when compared with larger companies. Staff numbers had to be comparatively larger while the line's gradients caused more wear and tear.

Keeling reported that the re-arrangement and signalling of the transfer and marshalling sidings at Lydney Junction was finished; one locomotive was being partly rebuilt at Bristol, a new one, *Gaveller*, ordered from the Vulcan Foundry at a cost of £2,085, and several coaches had been renewed.

In late 1890 and 1891 the Forest miners struck again, distress was as serious as it had been in the 1870s, stone breaking becoming the sole occupation of the miners.

On 27th February, 1891 it was announced that expenditure had increased considerably: four composite carriages had been bought from the Bristol Port Railway & Pier, one third-class coach, a passenger brake van and 12 mineral wagons. The £396 10*s.* 9*d.* spent on passenger carriages was charged to the reserve fund, but a new brake van was charged to the capital account.

Revenue had been £24,200 compared with £22,532, the greater part of the increase due to the carriage of minerals. The new engine, *Gaveller*, was expected to be delivered in March.

At the half-yearly meeting on 22nd August, 1891 Lucy revealed that total receipts were £23,486 compared with £22,733, but locomotive coal had cost £1,486 compared with £1,189, double the price of four years previously. Keeling said that the six ex-Bristol Port Railway & Pier coaches had been thoroughly renovated and proved very useful in carrying summer excursion traffic. *Gaveller* had proved satisfactory and the locomotive shed had been enlarged to provide an adequate fitting shop for repairs.

On 26th February, 1892 it was announced to the meeting that the upward trend in receipts had continued, being £24,200 compared with £23,150. Staffordshire coal was providing strong opposition as the gas works at Speech House was now supplied with coal from that source at a lower price than a nearby colliery was charging. That week two vessels at Sharpness were loading Staffordshire coal for Bridgwater and Ilfracombe. Although locomotive coal was twice the price and railwaymen's wages had increased, yet the price for carrying the coal remained fixed. The Forest colliers were to begin a strike on 12th March, 1892.

On 23rd August, 1892 the meeting heard that due to a fall of the Forest coal trade, gross receipts were £3,066 lower than the corresponding previous half-year, but expenses had been reduced by £1,877. The price of Forest coal was higher than other districts, even more than Scottish coal which had recently been shipped to Bridgwater and superseded Dean coal.

Keeling reported that additional sidings were being laid to the coal tip at the tidal basin and an iron footbridge had been erected across the line near Lydney church.

The meeting on 28th February, 1893 was told that a depression in trade had affected Forest of Dean coal. Traffic had fallen by 61,000 tons in the last two years and in turn railway receipts showed a decrease of £1,615. The reserve fund standing at £2,890 15*s.* 6*d.* needed a considerable addition to provide for much-needed renewals of rolling stock. Keeling reported that the Board of Trade had authorised the raising of further capital for the cost of interlocking points and signals.

Chapter Five

The GWR and MR Takeover

On 31st July, 1893 the following circular was sent to the company's debenture holders:

> I am instructed to inform you that in consequence of the long and severe depression in the Forest coal trade which became acute during the past half year, the Company's revenue for that period proves insufficient for the discharge of debenture interest. The prolonged labour difficulties in the Forest mining industry, which chiefly occasioned the depression referred to, have now resulted in the stoppage of the principal house coal collieries. An action has been taken against the Company by a holder of debenture stock preparatory to petitioning for the appointment of the directors as managers and the principal officers as receivers. Under the circumstances, the payment of interest must await the direction of the Court. You will readily appreciate the circumstances which have led to this result are entirely beyond the control of the Company, but it is hoped that the difficulties of the Forest district may soon be adjusted and the trade resume its normal condition. In this circular, it has been made abundantly clear that the Company is the victim of circumstances over which it has no control.
>
> The periodic traffic returns for months past show a serious decrease in the revenue of the Company and the stoppage of the principal house coal collieries, owing to the still existing dispute between the coal owners and the miners, has practically stopped the most important branches of the Company's traffic.
>
> We understand the Company's business will proceed as usual and that when trade is restored to its normal condition by a settlement of the strike, the directors hope to satisfy the claims of debenture holders, in which case, the Company would revert to the position it held prior to the recent unfortunate difficulty. Interest on some of the debentures, it may be added, is guaranteed by other Companies. [The Midland Railway and the Gloucester & Berkeley Navigation Companies paid the interest on the £75,000 4 per cent debenture stock.]
>
> The usual half-yearly meeting will shortly be held and the shareholders will then doubtless be informed as to the present financial position of the undertaking.
>
> Thomas Linton,
> Secretary to the Severn & Wye and Severn Bridge Railway Co.,
> Lydney, 31st July 1893.

GWR and Midland Severn & Wye Joint Railway third-class privilege ticket Cinderford-Sharpness.

At the meeting on 29th August, 1893 Lucy reported that mineral receipts had fallen from £16,395 to £13,675. Discontent among the Forest miners had been simmering since the start of the year and as the men worked less hours, railway receipts were reduced. The strike had meant that the SBR was unable to pay the debenture bond and under agreement with the guaranteeing companies, the MR, Sharpness New Docks and the Gloucester & Birmingham Navigation Company were called in to pay interest on the £75,000 four per cent debenture stock issued under their guarantee.

The engineer announced that the reduction in permanent way renewal had been balanced by more repairs to stations. A covered goods van and a brake van had been bought for £216 4*s.* 9*d.* Although some tasks had been deferred, essential work such as painting the Severn Bridge, carrying out repairs to the viaduct at Sharpness, improving the water supply to Lydney locomotive sheds had continued. Four and a half thousand tons of mud had been dredged from the Lydney Canal and a new house built for the harbour master.

Matters failed to improve and it was obvious that the only action was to sell the company even though this would be at a loss to most of the shareholders. At a board meeting in October 1893 George White, a major shareholder, revealed that in 1890 he had approached the Midland Railway to enquire if it would be interested in purchasing the Severn & Wye. The Midland declined as it would have cost £20,000 to raise the Severn & Wye line up to standard. White added that he had recently approached the

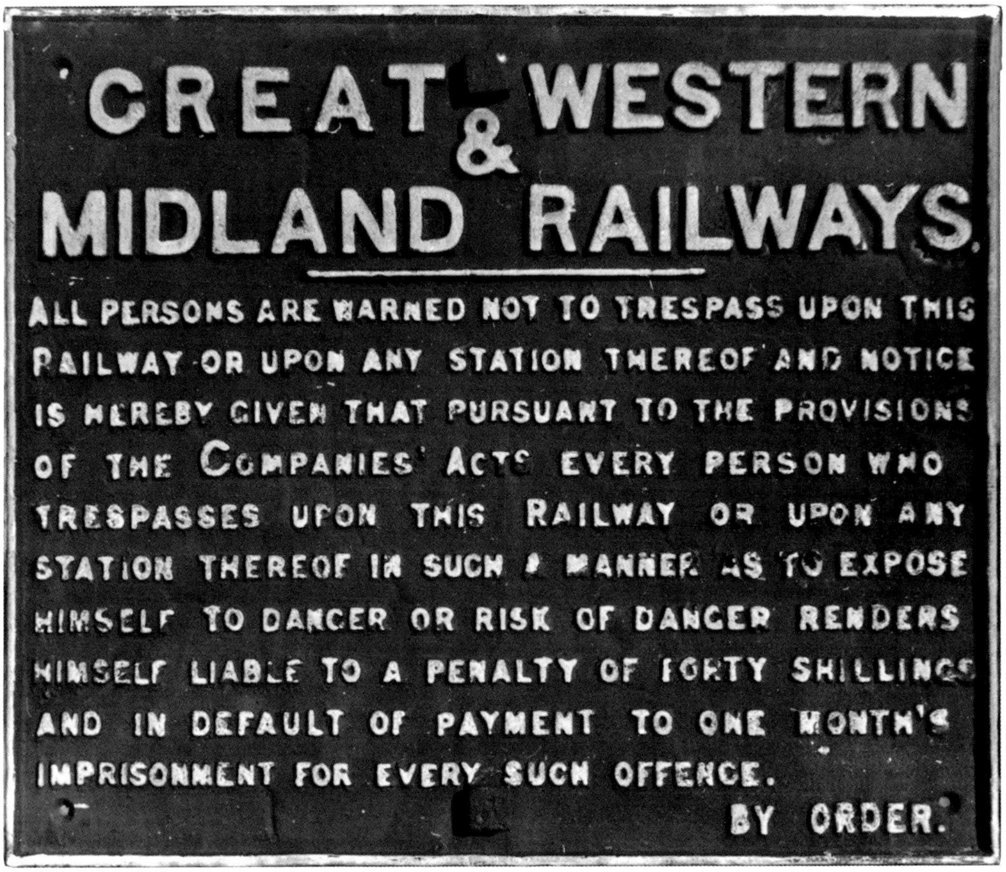

Great Western & Midland Railways cast-iron notice at Berkeley, September 1964. *D. Payne*

GWR, which then contacted the MR, with the result that a joint purchase would be feasible.

At a special meeting held on 27th February, 1894 Lucy said that the coal strike which started early in July 1893 had ended that September and since then traffic had been 'exceptionally good'. He said that there had been a considerable increase in the coal trade since September 1893 when the principal house coal collieries had resumed work. Passenger, parcels and mail traffic had brought in £2,819 compared with £3,058, but minerals and merchandise had earned £18,038 compared with £16,894 in the previous comparable half-year.

The annual tonnage carried in 1886 had been 622,804, but in 1893 only 537,695, so the company had failed to progress.

Lucy presented the terms on which Parliament would be asked to authorise the company's sale to the Midland and the Great Western. The price would be £477,300 for an undertaking which had cost £951,349. The sum received would be distributed:

To the registered proprietors of the £75,000 4 per cent guaranteed Debenture Stock, the sum of £125 for each £100.

To the holders of the £53,265 4 per cent Debenture Stock, £100 for each £100 of stock.

Subject to the payment of debts, the surplus was to be divided as near as possible in the following proportions:

To the proprietors of the £50,000 Preference A Stock, the sum of £53 for each £100 of that stock,

To the proprietors of the £107,467 Preference B Stock, £29 for each £100 of stock,

To the proprietors of the £298,269 Preference C 4 per cent Stock, £16 for each £100,

To the proprietors of the £167,348 Ordinary Stock, the sum of £12 for each £100 of stock.

Lucy said that the directors were of the opinion that these terms offered an equitable balance of the purchase money. The amounts at which the various classes were paid off would in every case exceed the market value of the respective stocks for the period 1885 to 30th June, 1893.

The date of the sale was to be 1st July, 1894 and it was agreed between the purchasing companies that the Midland Railway should on payment of a sum agreed on or fixed by arbitration, transfer to the GWR half its interest in the Berkeley branch which would then be worked jointly by the two companies as part of the Severn & Wye and Severn Bridge Railway. The GWR paid the MR £62,475.

The meeting unanimously approved the sale of the company and George White who had invested £75,804 in various classes of stock, said he proposed a resolution of thanks to the directors for the immense amount of time and trouble they had spent in carrying out the negotiations to a successful conclusion.

On 28th February, 1894 George Baker Keeling, (not to be confused with G. W. Keeling, engineer), died aged 80. He had joined the Severn & Wye Railway and Canal Company in 1847 as secretary and general manager and acted in this capacity until 1879 when he became managing director of the Severn & Wye and Severn Bridge Railway.

In the spring and summer of 1894 several objections were made to the sale, opposition coming from the Forest of Dean; Gloucestershire County Council; the Golden Valley Railway and the London & North Western Railway. The Bill received Royal Assent on 1st July, 1894 and so the Great Western & Midland Severn & Wye Joint Railway came into being.

The undertaking, known as the Great Western and Midland Severn & Wye Joint Railway (though interestingly after the 1923 Grouping the order was reversed to the London, Midland & Scottish and Great Western Railway Severn & Wye Joint Railway) would be controlled by a joint committee of the two companies. The GWR was to maintain the locomotives and supply coal, while the MR would tend the permanent way, signals, telegraph and weighing machines. Cash receipts and wages were shared, the MR controlling the section from Berkeley Road to Lydney and the GWR the rest of the Severn & Wye. A joint traffic inspector based at Lydney Town reported to the MR district controller and the Great Western divisional superintendent, both at Gloucester.

In addition to administering the Severn Bridge Railway and Severn & Wye Railway, the GWR & MR Joint Committee was also responsible for the Clifton Extension Railway, Halesowen Railway and the joint stations at Bristol, Churchdown, Worcester and Great Malvern. In the event, most problems were settled at the joint officers' quarterly meetings, only important questions being passed to the Joint Committee.

The 30th and final half-yearly meeting of the Severn Bridge Railway was held on 23rd August, 1894. W. C. Lucy in the chair reported that the receipts for the half-year were £20,610 13*s.* 5*d.* against £7,504 for the previous half-year, but expenditure had increased by £461. The net results were that the company had paid off £604 of the arrears of interest and added £110 to the reserve fund. An extraordinary meeting was then held to appoint liquidators, Messrs Lucy, Fripp, Keeling and Thomas Linton being selected to wind up the company's affairs. Sir William Marling moved a vote of thanks to Lucy for his long and valued service to the company and the last meeting of the company closed.

Lucy had fought hard for his company's financial success, but unfortunately had to contend with unstable conditions in the Forest and the competition of the Severn Tunnel. A public subscription was organised for a portrait of Lucy to be painted as he had lived in Gloucester for 44 years, been connected with the merchants of that city for most of that period including the Gloucester Banking Company and the Sharpness New Dock Company. On 4th May, 1895 Lucy was entertained to a dinner at the Guildhall, Gloucester, the mayor presenting his portrait painted by the Honourable John Collier.

The Joint Committee appointed John J. Petrie traffic manager on 12th August, 1895, who retained the post until 1898 when he transferred to the Midland & Great Northern Joint Railway. J. A. Carter succeeded him, but on Carter's retirement 31st December, 1919 the post was abolished, the MR and GWR supervising their respective sections from the MR district controller's office and that of the GWR divisional superintendent respectively, both located at Gloucester. A joint traffic inspector at Lydney Town then dealt with the arrangement of trains, extra workings and staff matters.

In 1895 the GWR proposed building a 14-mile long line from the planned Bristol & South Wales Direct Railway at Chipping Sodbury, to join the Severn & Wye Railway between Berkeley Road and Sharpness. In 1896 this scheme was withdrawn on the opposition of the MR, the GWR substituting a more economical short spur from its Badminton line to join the MR at Yate and making another spur south of Berkeley Road station to connect with the joint railway, the GWR exercising its running powers acquired in 1846 over the intervening 11 miles of MR track. The Berkeley loop was constructed for the GWR by Charles Baker & Sons and to assist the work, this contractor purchased a new Manning Wardle 0-6-0ST Works No. 1642 naming it *Frank*. The Westerleigh and Berkeley loops were both opened for goods traffic on 9th March, 1908.

The Severn Bridge Railway had suffered no serious accident, but two minor mishaps occurred during the early years of the joint companies. On 27th May, 1896 about 7.30 pm the brake van of the 7.20 pm goods

Sharpness-Lydney came off the road at the catch point near the Severn Bridge station. The train had come to a stand just clear of the point, but the guard had omitted to apply his brake and the couplings being slack allowed the van to run back and become derailed.

On 5th June, 1897 the permanent way inspector reported: The 3.45 pm passenger train in crossing from the North Branch to the Down road through the Up goods road, ran through facing points near Sharpness station causing the switch tongue to bend.

As its early critics anticipated, although its promoters claimed otherwise, the bridge did indeed prove an impediment to shipping, the 39-ton trow *Brothers* colliding with a pier and being wrecked in 1879. On 9th October, 1904 a trow from Chalford was conveying a load of road stone from Chepstow to Framilode, despite the exertions of the crew, was carried by the treacherous current against No. 13 pillar and sank almost immediately. Fortunately the owner, captain and a sailor had climbed into a small boat and this saved their lives. On the tide receding the damage was patched and the vessel towed to Chalford.

Although railway enthusiasts love locomotives, they do not always appreciate that most horses regard them with considerable suspicion. On 18th October, 1904 R. A. Lister, JP and prospective candidate for the Tewkesbury Division, in his JP capacity had driven to Berkeley to sign an extension of time for a licence. On his return, near Berkeley Road station, his horse became frightened by a train on the Severn & Wye Joint Railway, dashed into a wagon, violently throwing out Lister and his coachman. Lister turned a somersault and was severely shaken, but his coachman, badly injured, was taken to Dursley. The coach was smashed, and despite what must have been a considerable shock, Lister was able to attend Gloucester Quarter Sessions that afternoon.

On 1st January, 1906 the line from Berkeley Road to 6 chains south of Coleford Junction was transferred to the MR for maintenance purposes, the GWR responsible for the rest of the joint system.

Until about 1927 as oil tank wagons were prohibited from passing through the Severn Tunnel they were diverted over the Severn Bridge.

In the 1930s the GWR attempted to have weight restrictions over the bridge eased so that it could use larger and more economic locomotives, but the joint owners, the London Midland & Scottish Railway which was responsible for its maintenance, rejected the idea. In 1932 a scheme was proposed to raise the bridge decking to rail level to enable the bridge to be used for road traffic in addition to that by rail. The idea was not proceeded with, but resurrected in 1939. It was then abandoned due to the problems of separating road and rail traffic and the perils of strong cross-winds blowing traffic into the bridge sides.

One of the most serious bridge strikes occurred on 4th February, 1939, involving three petrol-carrying tanker barges, the *Severn Traveller*, *Severn Carrier* and *Severn Pioneer*, all owned by the Severn & Canal Carrying Company Limited of Gloucester and Bristol. *Traveller* and *Carrier* each had a crew of three, but *Pioneer* had just two. They were steel, flat-bottomed barges built for plying between Avonmouth and Worcester. *Traveller* and *Carrier* were motor-propelled but *Pioneer* a dumb barge.

Traveller led, and roped to her was *Carrier*, with *Pioneer* bringing up the rear. A single barge was insufficiently strong to haul *Pioneer* which was why they were 'double-heading'.

They left Avonmouth at 5.15 pm and shortly after 7.00 pm were abreast of Sharpness. As *Traveller* swung round to enter the docks, *Carrier* turned turtle, *Traveller* collided with her and heeled over, *Pioneer* colliding with the Severn *Bridge*.

Captain Albert Tonks on *Traveller* was thrown into the water, but as his vessel righted herself, he clung to her and pulled Frederick Vincent, his engineer, onto the deck. Despite the anchor being dropped, *Traveller* drifted with the tide.

While this drama was taking place, Captain G. T. Owen, the Sharpness harbour master, ordered the tug *Primrose* lying in the tidal basin, to put out. To save time she was manned by a scratch crew.

Near the bridge they found *Pioneer* very low in the water and her crew missing. The tug *Primrose* secured *Pioneer* and drew her to Sharpness where the fire float pumped her dry and as soon as Captain Owen deemed her safe, she was moored by the lock.

Meanwhile, *Carrier* and *Traveller* were swept up river in the darkness, *Carrier* with no living soul aboard.

As Lionel Keedwell, aged 22, son of the proprietor of the Berkeley Arms Inn on the river bank at Purton was going to bed, he saw lights flashing on the river and realised that someone was in difficulty. Shouting from the bank he received a faint reply. Braving the treacherous currents he collected five local men, launched his boat and found Captain Tonks and Vincent on the *Traveller*. Tonks had received a face wound, was given care at the inn and by next day had almost recovered.

Daylight saw *Traveller*, slightly damaged, held by anchor on a sandbank about half a mile above the bridge, while *Carrier* was about two miles further up but started to drift downstream with the ebb tide, coming to rest about 1,000 yards above the bridge.

The bridge was closed for 24 hours in order that it could be checked for possible damage, but none was found.

Two bodies found near Purton were brought in by boat; the watch on one corpse had stopped at 7.10. The third body was spotted in the afternoon and eight men set off with a ladder to retrieve it. In places the mud was so soft that carrying the ladder used as a stretcher, they sank so deeply that the ladder and its load had to be dragged along with ropes.

Greater details were revealed when the inquest on the three bodies was opened at Berkeley Police Court on 8th February, 1939.

Albert Tonks, aged 29, was responsible for the tow. The coroner warned him that he was not obliged to give evidence, but said he was willing. He observed that going with the tide had given them a good trip and they arrived outside the pier at Sharpness about 7.05 pm. Standard procedure for entering the harbour was that they slewed round and 'put the head into the tide and pulled back into the entrance of the dock'. He had entered in that manner many times.

They started to go into the docks under their own power, *Traveller* and *Carrier* using their engines. The first sign which made Tonks aware of anything untoward was when Walter Capener, aged 18, the third hand called 'The rope is slack!' Tonks said:

A dolphin protecting the columns supporting span No. 19 and No. 20, 1956. *BR WR*

> We hauled the rope in. It was entirely disconnected. Then we knew that the other boat [*Carrier*] was drifting.

The other barge was not strong enough to haul the dumb barge by itself. Tonks said that he circled round and came alongside *Carrier*, threw her a tow rope and made it fast.

By this time the vessels had gone about half a mile beyond the dock entrance and were near the bridge.

> We circled round with the rope taut, *Pioneer* collided with the bridge and sank. *Carrier* was also carried back and collided with another bridge pier, partly sinking, her stern below water and her bow came up.

Traveller collided with the same bridge pier as *Pioneer*, heeling right over and this was when Capener went overboard. The propeller was not holding as they were going with the tide. They dropped anchor but it dragged, and when it finally held they were on the Gloucester side of the bridge. He tried to signal ashore by flashing lights. About two to two and a half hours later on the ebb tide they settled on the sand. No one came to assist until the tide went down.

The primary cause of the disaster was the rope becoming detached. It was the *Carrier's* end of the rope which came loose, the 5½ inch thick rope came adrift, but did not break. The rope was spliced with an eye at each end and when it came adrift, both eyes remained intact.

Frederick Vincent, engineer, said that he came up from the engine room and gave Capener a hand at hauling in the tow rope and they were turning round to get alongside the *Carrier*. Joubert Matthews, aged 20, engineer on *Carrier*, told Vincent that the rope had slipped off, but gave no reason. Vincent said they had started taking the strain again between the dock entrance and the bridge and were heading to Sharpness. Next thing *Pioneer* struck the bridge and both engines were working at that time. When the tow rope was made fast *Pioneer* was about 300 yards from the bridge, but the engines were not making headway against the tide, so they were swept towards the bridge. Vincent said when *Carrier* struck the bridge he slipped the tow rope. They collided almost broadside and heeled over. Thrown into the water, he grabbed one of the life chains. He was in the water for about a minute. *Traveller* righted herself and he got aboard. Tonks was not completely thrown off.

The speed of the tide increases above Sharpness Docks and this causes the Bore; the Severn is navigable below but dangerous above and it was surprising that no qualification was necessary to become a bargemaster. The final inquiry was adjourned until 28th February, 1939 in case more bodies turned up. The coroner commented on the fact that no proper stretcher was available at Purton and men had to use a ladder. He suggested the matter be brought to the attention of the local council. A fund was started to aid the dependants of the six bargemen drowned.

In October 1943 Pier No. 17 was struck by a dumb barge carrying 400 tons of grain, the impact so serious that it sheared off six of the bolts which held the fixed bearings to the topmost castings of the cylinders, shattering the cast-iron bracing framework between the two cylinders from top to bottom. LMS engineers repaired the pier and substituted mild steel bracings strapped and bolted to the cylinders. It was significant that in the report no mention was made of any fracturing of the cast-iron at the base of the cylinders or of any disturbance to the foundations, although the blow could have been sufficient to have cracked the cast-iron around the bases of both cylinders.

The bridge survived the Second World War, being narrowly missed by a bomb which fell on the road near Purton Manor. During the same period, many daring young RAF pilots on training flights in Spitfires and Hurricanes gained a thrill passing below one of the 312 feet spans and on one occasion a Vickers Wellington bomber was seen to do this. An onlooker admired these pilots until he saw men painting the bridge, hanging in their cradle, while an aircraft few within a few feet of them. Following complaints, a detachment of RAF police was billeted at the Severn Bridge Hotel and from a platelayer's hut near the viaduct they maintained a vigil recording aircraft numbers and after a few courts-martial this daredevil practice ceased.

Following Nationalisation, on 9th February, 1948 the commercial and operating responsibility for the line was transferred to the Western Region, while from 5th April, 1948 signal and telecommunications as far as Sharpness was in the hands of the London Midland Region. Although the bridge had been maintained by the MR, then the LMS and then the London Midland Region, on 11th July, 1948 it became the entire responsibility of the Western Region.

At noon on 22th April, 1951 Class '4F' 0-6-0 No. 44266 of 22A, Bristol (Barrow Road) derailed under the road bridge on the Lydney branch near Berkeley Road. Crane RS 1062/20 and Class '4F' 0-6-0 No. 44055 of 22B, Gloucester (Barnwood), were in attendance.

On 27th February, 1957 M. Price Philips, MP for Gloucestershire West asked Harold Wilkinson, the Minister of Transport, whether in view of the postponement of building the Severn Road Bridge and the consequent traffic congestion at Chepstow, whether he would consult with the British Transport Commission with the view to converting the railway bridge to take road traffic as well. He replied that the proposal had been examined many years before and also in 1956, but as only extremely restricted road use would be possible it would not justify the expenditure involved.

In the 1950s it was still the practice for wagon-loads to be sent to the goods yards of local stations. This method of working required shunting to take place: re-marshalling wagons from pick-up freights into long-distance trains for various destinations and then probable further shunting to form a local goods train to take them to their final destination. In the 1950s it was realised that shunting would be more efficient at new marshalling yards using gravity and one such yard was to be laid at Brookthorpe, near Gloucester. If this came about, it was thought that more use could be made of the Severn Bridge route as it offered a short cut to South Wales. The problem was that it had been classified a 'Yellow' route, (the GWR classified an engine's weight by colour) and only relatively light locomotives such as 0-4-2Ts, 0-6-0s, 0-6-0STs and 0-6-0PTs were permitted to cross. 'Blue' engines of the '63XX' 2-6-0 and Manor class 4-6-0s were allowed to cross with special permission. Had the Brookthorpe

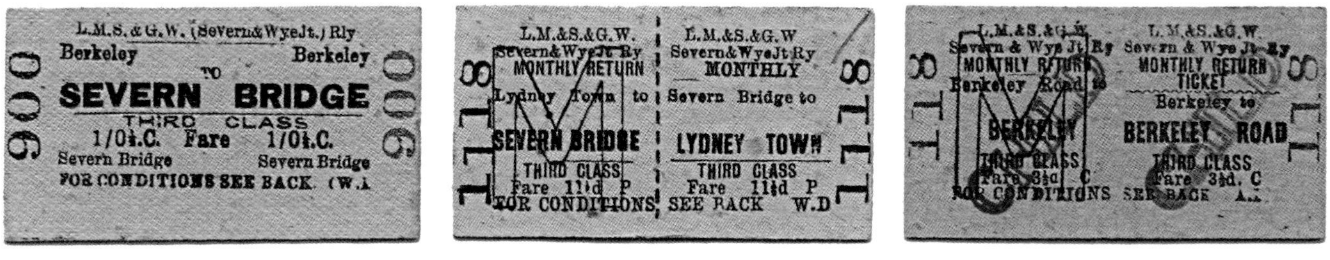

Selection of tickets.

scheme come to fruition, it would have been advantageous if 'Red' engines, that is all ex-GWR locomotives except the 'King' class which were double-red, could have run over the bridge.

Calculations suggested that the bridge was capable of carrying Red locomotives at a speed of three revolutions per second without any strengthening to the girder work. The chief civil engineer of the Western Region thought it prudent to test these figures practically. The research department of the British Transport Commission attached electric resistance strain gauges at various places on the bridge to record the changes in strain during the passage of a test train.

On 18th April, 1956 members of the British Transport Commission's research department met members of the district civil engineer's office at Gloucester and decided that as bridge painting would start that coming Whitsun, the painters could assist in erecting any scaffolding for the strain gauges when and where it was required. This proved a difficult operation as to receive the gauges several layers of paint had to be removed and then a smooth surface prepared on the exposed metal. The gauge recording equipment and power supply was placed in one of the shelters formed by the finials at the western end of the bridge and to avoid long cable runs another base was set up at the swing bridge. Fifty-seven straining gauges were attached to three spans of the bridge: 19 on the 134 feet span; 20 on the adjoining 312 feet span and 18 on the swing bridge.

The dates for the tests were Sundays 15th and 22nd July, 1956. The test train consisted of two 'Castle' class 4-6-0s hauling a train of eight 20-ton capacity Grampus mineral wagons loaded with ballast plus a brake van. The drivers were issued with the instructions:

> Under no circumstances are the two locomotives to be allowed to proceed into the station loop at Sharpness or proceed into the North Docks branch. Sufficient water to be taken on at Lydney for the day's operations. The test train must be propelled tender-first over the Bridge.

On 15th July, 1956 No. 5018 *St Mawes Castle* and No. 5014 *Winchester Castle* left their home shed, Gloucester, for Lydney to form the heaviest test load since 3rd & 4th October, 1879. The test train of approximately 480 tons arrived at the bridge soon after 8.00 am.

Several runs at 5 mph were made over the 312 feet span, a deflection of 1½ inches being measured – the same as Colonel Rich found in 1879. Tests were then made over the 134 feet span using just the two locomotives coupled together.

On 22nd July, 1956 the swing span was tested using the same engines. As analysis revealed that on 15th July gauges on the 134 feet span had not functioned correctly, so on 1st August, 1956 between train occupations, a further series of test runs were made just using No. 5018, No. 5042 being unavailable.

Although the stresses were generally low, it was believed that the high stress in the diagonals in conjunction with the vibrations would prohibit the use of the 'Castle' class. The danger was fatigue. As the wrought-iron used on the bridge was similar to that used in Over Bridge at Gloucester and the Old Clyde Bridge, Glasgow, specimens from these bridges, including a rivet hole, were tested and offered a fatigue limit stress range of 7 tons per square inch. A table of stress and frequency was compiled:

Locomotive type	Stress range tons per square inch	Frequency of crossing per year
'Castle'	12	?
'63XX' 2-6-0	9.6	50
'2251' 0-6-0	7.6	200
'16XX' 0-6-0PT	4.7	700
'2021' 0-6-0PT	3.5	1,300
'48XX' 0-4-2T	4.0	3,000

Conclusions were that all stresses of 7.0 tons per square inch and below were considered as making no contribution to cumulative damage. The frequency at which it was proposed to run 'Castle' class locomotives over the bridge was unknown, but at four times a week, life for the bracings would be 14 years. Although the remaining life of these stays seemed short, the probability of failure was only 1 in 100 so that in the time stated, only one bracing member in approximately three spans would fail. As the failure of one diagonal would not have serious consequences provided that a regular and thorough inspection of the bridge was carried out, a very limited use by engines of the 'Castle' class might be permitted.

If the bridge was to be kept in service beyond 1970 the double system of ties or tension diagonals would be replaced by diagonal web members carrying primary compressive stresses This would permit Red engines to use the bridge at restricted speed for at least 50 years. As sufficient skilled steel erectors were not available in BR's own workforce, tenders were invited for replacing 488 diagonals on all spans except the swing portion and requiring the fitting of approximately 50,000 bolts. Badly grooved roller suspension bearings were to be replaced with rubber-cum-steel sandwich blocks.

In 1959 the report of the British Transport Commission to the Ministry of Transport on the re-appraisal of the plan for modernising BR said that the Severn Bridge route would be developed to relieve the congested Severn Tunnel.

The Fairfield Bridge & Shipbuilding Company, Chepstow won the £95,000 contract for strengthening the bridge, work commencing 4th April, 1960. By the end of September 1960 the first small span and the two 312 feet spans had been modified and the scaffolding erected on span No. 18.

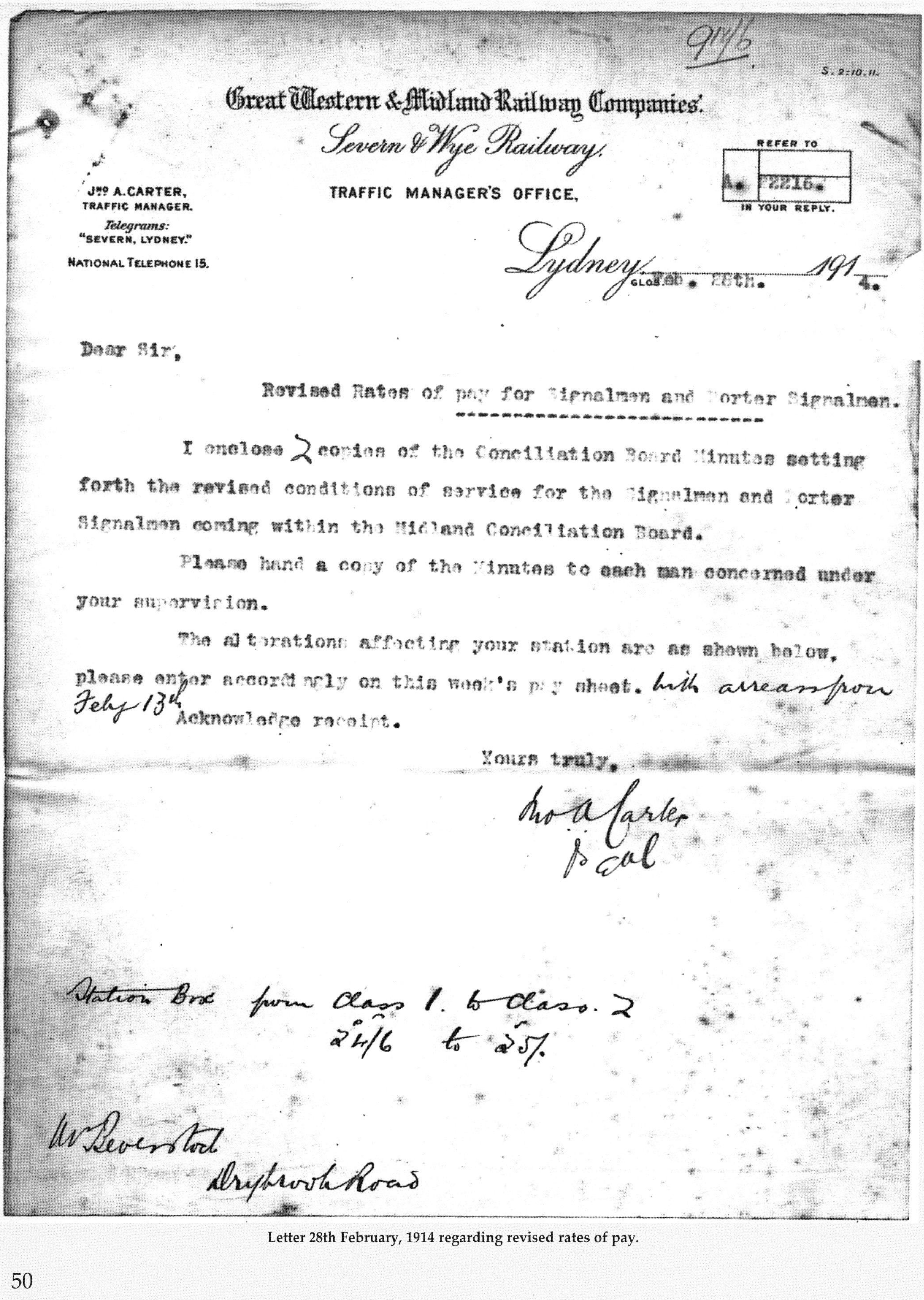

9146

S. 2:10.11.

Great Western & Midland Railway Companies.

Severn & Wye Railway.

REFER TO A. 22216. IN YOUR REPLY.

JNO A. CARTER,
TRAFFIC MANAGER.
Telegrams: "SEVERN, LYDNEY."
NATIONAL TELEPHONE 15.

TRAFFIC MANAGER'S OFFICE,

Lydney, GLOS. Feb. 28th. 1914.

Dear Sir,

Revised Rates of pay for Signalmen and Porter Signalmen.

I enclose 2 copies of the Conciliation Board Minutes setting forth the revised conditions of service for the Signalmen and Porter Signalmen coming within the Midland Conciliation Board.

Please hand a copy of the Minutes to each man concerned under your supervision.

The alterations affecting your station are as shown below, please enter accordingly on this week's pay sheet. with arrears from Feby 13th.

Acknowledge receipt.

Yours truly,

Jno A Carter
p C.O.L.

Station Box from Class 1. to Class. 2
24/6 to 25/.

Mr Beverstock
Drybrook Road

Letter 28th February, 1914 regarding revised rates of pay.

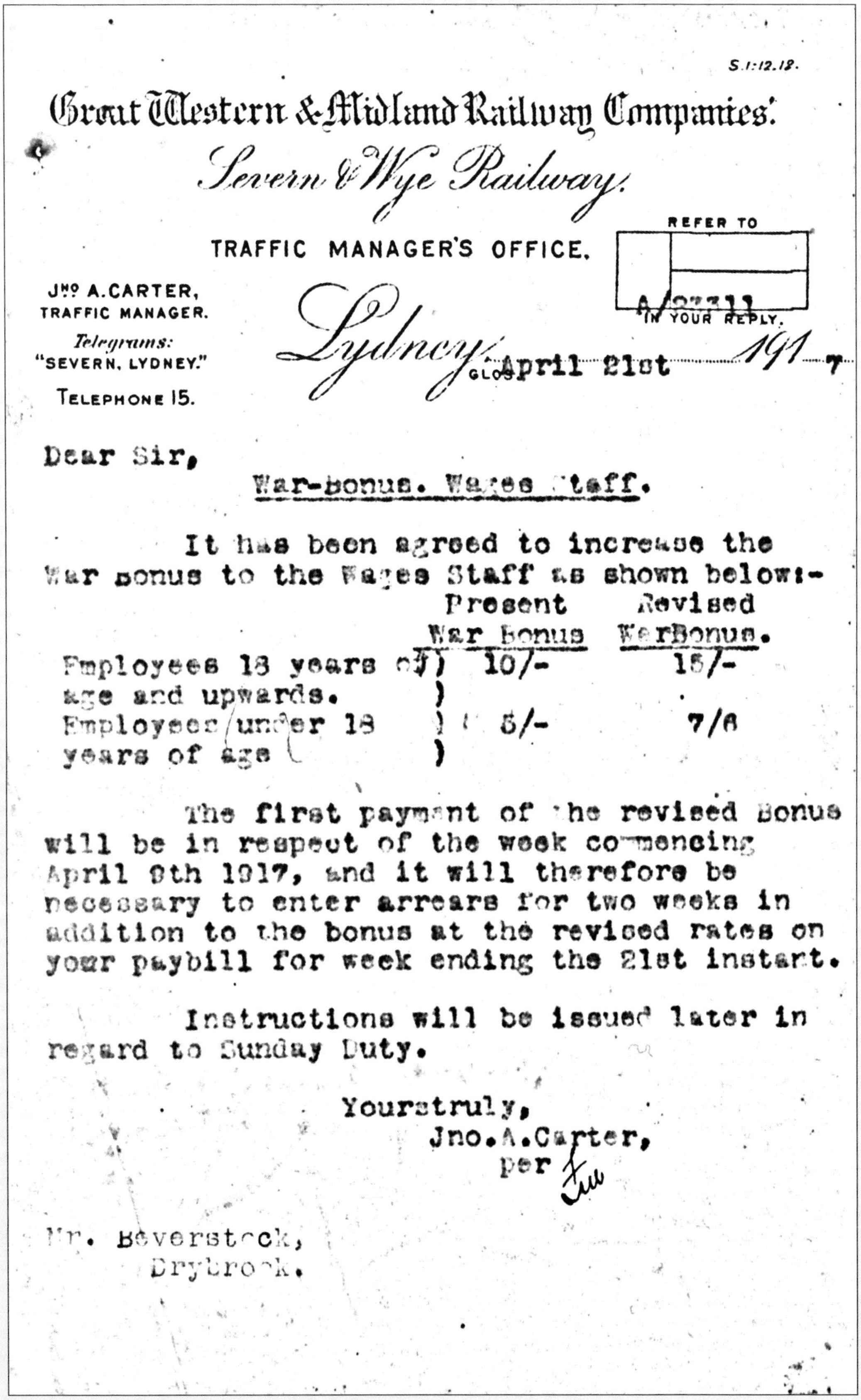

S.1:12.18.

Great Western & Midland Railway Companies.

Severn & Wye Railway.

TRAFFIC MANAGER'S OFFICE.

REFER TO A/27711 IN YOUR REPLY.

JNO A. CARTER, TRAFFIC MANAGER.

Telegrams: "SEVERN, LYDNEY."

TELEPHONE 15.

Lydney, GLOS April 21st 1917

Dear Sir,

War-Bonus. Wages Staff.

It has been agreed to increase the War Bonus to the Wages Staff as shown below:-

	Present War Bonus	Revised War Bonus.
Employees 18 years of age and upwards.	10/-	15/-
Employees under 18 years of age	5/-	7/6

The first payment of the revised Bonus will be in respect of the week commencing April 8th 1917, and it will therefore be necessary to enter arrears for two weeks in addition to the bonus at the revised rates on your paybill for week ending the 21st instant.

Instructions will be issued later in regard to Sunday Duty.

Yours truly,
Jno. A. Carter,
per

Mr. Beverstock,
Drybrook.

Letter regarding War Bonus, 21st April, 1917.

Chapter Six

Facing page: View across the gap from span No. 16 to No. 18, 27th October, 1960.
BR WR

Two Bridge Spans Destroyed: The Beginning of the End

At 10.16 pm on 25th October, 1960 signalman Donald Dobbs cleared signals for the 9.45 pm Lydney-Stoke Gifford freight hauled by '43XX' class 2-6-0 No. 6394. As it passed over the bridge Driver Donald Powell and Fireman Cliff Reeks found it eerie as the fog prevented them from seeing the water. After the train had cleared the section, at 10.30 pm T. C. Francis, the new works inspector, left the Severn Bridge station signal box after collecting the keys to permit the night occupation of the bridge.

Concurrently a group of 16 ships on the incoming tide was proceeding upstream from Avonmouth to Sharpness and in places the fog was quite dense.

As Mr Francis walked along the Up platform he saw a sheet of red flame rise higher than the bridge followed by an explosion. He raced to the bridge, saw two tankers ablaze and rushed back to the signal box to use the telephone to notify the ambulance service and the police. He returned to the bridge, walked to span No. 18 and was horrified to discover that Nos. 17 and 16 were completely missing.

John Harker Limited of Knottingly, Yorkshire owned a fleet of esturial tanker barges, its vessels named after Yorkshire Dales. One was the MV *Arkendale H* built as a dumb barge tanker in 1937 and converted to motor-power in 1948. Another was the MV *Wastdale H* built in 1951

View west from span No. 19, 27th October, 1960.
BR WR

Captain George Thompson, skipper of the MV *Arkendale H*, carrying a load of 296 tons of fuel oil and having a total weight of 408 tons, stated that in the dense fog near the entrance to the docks, the MV *Wastdale H* loaded with 351 tons of motor spirit and having a total weight of 450 tons, drifted quietly to his vessel and became locked by suction, making it difficult to separate them. Neither master realised that they had drifted past the lock entrance at Sharpness – they had intended using the Gloucester-Sharpness Canal to Gloucester. Locked together and travelling sideways at 3½ – 4 knots, *Wastdale* struck Pier 17 – the one which had been damaged in October 1943. She turned over on her port side, the tide thrusting *Arkendale* on top.

The two 174 feet spans supported by the pier collapsed on to the barges, the tide carrying both upstream to ground on a sandbank.

Captain Thompson of *Arkendale* saw two of his crew and knowing that they were unable to swim, threw each a lifebelt and told them to jump into the water with him. The river for the ¾-mile width and for a mile upstream was ablaze with petrol; he jumped but the other two did not.

Captain James Drew of the *Wastdale* saw two of his crew inflate a life raft, but drift away as it was thrown into the water. He led them into the water. His engineer, Jack Cooper, swam below the burning oil.

The current took Captain Thompson three miles upstream before he was able to land on the west bank. Captain Drew, wearing a life jacket also landed on the west bank. Both were taken to Lydney Hospital in the same ambulance.

At Sharpness, fire service personnel loaded a small rowing boat on a lorry where it was taken two miles upstream to Purton and carried over the marsh to the water. At great risk to their own lives they set out to look for survivors. Hearing shouts they found engineer Jack Cooper wearing a life jacket and exhausted. He, too, was taken to Lydney hospital. A total of five men lost their lives.

Heat from the fire was sufficiently intense to clear the fog and enabled the remaining vessels out in the river to gain a safe entry to the lock at Sharpness.

An alert employee at the Sharpness gas holder immediately shut off the supply before any gas could ignite. Although Sharpness had a fire-boat this was only suitable for dock or canal use and it would have been unwise to have taken it out on the river especially under the prevailing conditions. Signalman D. C. Dolman said that the impact and explosion shook the whole bridge violently. Burning fiercely, the two tankers went aground about half a mile on the upstream side of the bridge midway between Sharpness and Purton.

The columns between span No. 19 and No. 20, 15th July, 1956.

BR WR

The 30 plus Fairfield's men working on strengthening the two spans which fell, owed their lives to the fact that the BBC Light Programme was broadcasting a live commentary on a title fight between the French boxer Alphonese Halimi and Freddie Gilroy from Belfast, for the vacant World Bantamweight Championship at the Empire Pool, Wembley. That evening the bridgemen took their break to coincide with the broadcast. This meant that when the tankers struck the bridge they were not, as normal, out working, but safely in Severn Bridge station listening to the radio. The outcome of the fight? The Frenchman won on points.

Daylight revealed that Pier No. 17 and two spans had completely disappeared and the 12 inch gas main from Dursley to the Forest of Dean, only installed in 1954, was severed, so many homes had a cold breakfast that morning. The South Western Gas Board had to supply butane cylinders, electric fires and so forth to Forest consumers until 8th November, 1960 when a four-inch diameter plastic pipe was slung across the gap. In due course a new pipeline was laid across the river further up. GPO communications via the bridge were also cut.

That morning the Sharpness – Lydney school train of necessity ran via Gloucester and was worked by Class '4F' 0-6-0 No. 44167 and as it travelled tender-first, its crew had an unpleasant experience. In the ensuing days, the 50 or so Lydney Grammar School pupils from Berkeley and Sharpness continued to travel by train, but via Gloucester, involving a daily return journey of almost 80 miles amounting to nearly 400 miles each week. It necessitated three trains and a bus ride to reach a school only three to four miles away from their homes.

They were required to be at Sharpness station for the 7.55 am, an auto train of one coach leaving Berkeley at 8.01. At Berkeley Road they changed to a Bristol to Bradford express which made an unadvertised stop at 8.18 am arriving at Gloucester Eastgate 8.41. They strode quickly across the 250

A Down auto train propelled by '14XX' class 0-4-2T No. 1472, 22nd August, 1964. The disused Down platform is on the left. Notice that due to shorter trains, the platform has been curtailed. *W. Potter*

yard long covered footbridge to the Central station to catch the 8.50 all stations to Cardiff, usually worked by a Gloucester or Swindon-built Cross-Country DMU which arrived at Lydney Junction 9.08 am.

The final mile was in the school bus operated by Willett's of Yorkley, but reaching it entailed care crossing the South Wales main line on a boarded foot-crossing, usually just in front of their DMU still at the platform and with a Cardiff to Newcastle express approaching on the Up line. Lydney Grammar School was reached at 9.25 just at the end of morning assembly.

Their return journey required them to miss the final lesson of the day as the bus left the school gates at 3.20 pm. The two brake-third, compartment-type auto trailers were usually drawn by a '54XX' or '64XX' auto-fitted 0-6-0PT, though sometimes a '14XX' 0-4-2T was rostered. Leaving Lydney Junction at 3.30, 26 minutes were allowed for the 19½ miles to Gloucester Central where it stopped on the middle road to take on water.

At 4.01 it left for Berkeley Road where it stopped to set down some pupils before arriving at Berkeley 4.32 and Sharpness 4.37. The season ticket charge for this mileage was less than two shillings a day. At Sharpness the crew topped up the tanks and made two further trips to Berkeley Road before leaving Sharpness as empty coaching stock at 6.55 pm and reaching Lydney Junction at 8. 40. The train was propelled the 38½ miles Sharpness to Lydney Junction and currently this was probably the longest journey a steam locomotive pushed its train. From 9th September, 1962 this school service was withdrawn and pupils enjoyed a mere seven mile bus journey to Dursley Grammar School.

On 26th October, 1960 the army placed explosives in the stern and bows of both tankers so that they would sink into the mud to prevent their movement in the river; they remain there today. In due course a new gas pipeline was laid across the river further up.

The two tankers which destroyed the bridge seen here beached.
Author's collection

On 27th October, 1960 the auto train service resumed running just from Berkeley Road to Sharpness, until the branch closed to passengers 7th September, 1964. Despite the loss of the spans, as 18 had been already strengthened, work continued on the Sharpness side on 7th November, 1960, the contractors having to carry up to 15 men daily in a Bedford van between Chepstow and Sharpness via Gloucester. At the time it seemed that the chance of getting on with the work while the bridge was closed to traffic was simply too good to miss.

In November 1960 the chief civil engineer's department, Western Region, provided drawings for the bridge's restoration, the cost of which was estimated to be £188,000. The missing pier would be replaced by a new concrete pier, while the damaged pier would be repaired. The two missing spans would be replaced by a single, welded mild steel span supported in the centre by the new concrete pier.

During 1961 the case between the ship owner's John Harker Limited and the British Transport Commission, the Fairfield Bridge & Shipbuilding Company Limited, the South Western Gas Board and the Postmaster General came before the High Court of Justice, Admiralty Division.

Had the bridge been wrecked by fire, lightning or explosion, BR could have expected a cost of up to £1½ million to be met, but examination of the girders showed that the vessels struck before the explosion. Unfortunately under the 1956 Merchant Shipping Act the limited liability for damage through collision did not exceed the sum equal to about 24 times the net registered tonnage of the vessel, or vessels. BR only received £5,000, while Fairfield, which lost plant to the value of £10,000, received just a little over £100. In May 1961 the cost of repairing the bridge was estimated to be £294,000 and would take 18 months to two years to complete. As it was intended that the fallen spans were to be replaced, Messrs Fairfield continued the strengthening work and completed it on 27th October, 1961. Meanwhile on 17th February, 1961 the MV *BP Explorer* capsized below Sharpness drowning its entire crew of five, drifted upstream under the bridge, and then

The Severn Bridge with the South Wales line in the foreground, 9th August, 1961. *Author*

struck Pier No. 20 when drifting downstream on the falling tide. Bridge damage was estimated to be another £12,740. The tanker was repaired but later lost at Barry.

Early in December 1961 the MV *Universal Dipper* owned by the Universal Divers Limited, Liverpool, carried out an extensive underwater survey in the area around Piers Nos. 16, 17 and 18. It was decided to erect a temporary trestle under the bridge next to Pier No. 16 which leaned slightly towards the Sharpness bank. A contract was won by Peter Lind & Company Limited to commence work on 31st January, 1962. Gales in the Irish Sea delayed the arrival of the twin floating crane hired by Peter Lind.

Then on 14th April, 1962 the twin-hulled floating crane *Tweedledum and Tweedledee* hired by Peter Lind & Company Limited on a contract to erect a temporary trestle to support Pier No. 16 which was leaning so severely it was in danger of collapse, broke from its moorings on the flood tide. The MV *Magpie* made several abortive attempts to take it in tow until *Magpie's* propeller became entangled in a rope. The crew watched helplessly as the crane drifted downstream, smashed against the dolphins on Pier No. 20 and its jib struck the underside of the bridge causing £6,000 damage before the crane finally ran ashore off Lydney. Following repairs at Avonmouth it arrived back on station 7th May, 1962 – the contract was supposed to have been completed by 2nd May!

Despite these problems the possibility of repairing the bridge was still pursued, the estimated cost of which had risen to £300,000 plus £600 for repairs to Pier No. 20; the alternative of demolition was estimated to have cost £250,000.

BR took more than six years to make up its mind what to do with the bridge – one option pursued was that of selling it to the Central Electricity Generating Board to carry high-voltage cables across the river, but eventually concluded that the only hardship caused by the route's closure and the bridge's demolition would be when on winter Sundays when the Severn Tunnel was closed for maintenance and it would add half an hour or so having to make the detour via Gloucester.

In 1967 the Nordman Construction Company of Gloucester proposed a choice of four schemes to dismantle and remove the bridge down to river bed level, the original idea being to recycle spans to form bridges in South Wales and Spain.

Scheme 1
: To jack up the trusses, replace the old bearings with new roller bearings, strengthen the trusses with continuous wire cables and draw the spans in trains of five at a time to the end by winches.

Scheme 2
: To erect a derrick over the piers; lift, swing outwards and lower each truss to the river bed at a suitable state of the tide. Winch each truss, previously fitted with skids, to shallow water.

Scheme 3
: To cut out the floor and top bracing; move the trusses together; erect a derrick as in Scheme 2 to take the weight of the unsupported ends and draw the girders back between the trusses of the next and each succeeding span.

Scheme 4
: A floating crane to lift each span complete (except for the two 312 feet spans) and deposit them on each side of the river for cutting up. This was the scheme adopted.

Ulrich Harms floating crane *Magnus II* was towed from Hamburg, but for manoeuvring possessed its own diesel-electric propulsion with steering screws at each corner of its hull. To counterbalance the weight of the jib when lifting, water was pumped into ballast tanks at the stern, thus allowing it to maintain an even keel. Its lifting capacity of 400 tons to a height of 150 feet over a radius of 80 feet made it capable of dealing with all the bridge spans except the two 312 feet spans over the navigation channel which each weighed 533 tons.

Magnus II arrived on 21st August, 1967. Work started the following day and the three 174 feet mid-stream spans each weighing 170 tons were deposited on the Lydney foreshore. It was intended to use a specially-designed cradle, but the crane master, under whose direction the lifting was carried out, was satisfied that a sling would suffice. Other spans were

Magnus II floating crane used to lift the spans, August 1967.
Author's collection

Removed span placed below the viaduct, 24th August, 1967.

BR WR

removed by the floating crane placed on the east bank for cutting up. As the two 312 feet long spans over the navigation channel were in excess of the crane's capacity, these were freed from their bearings and drawn by hawsers attached to the crane until together with the pier they dropped into the river. The intention was that a diver would use an oxy-acetylene burner to cut the metal into suitable weights to be lifted by the crane into a barge. Unfortunately every attempt to lift the metal failed due to the hook slipping off.

At this point the river authority became concerned with the obstruction caused by the mass of metal in the navigation channel and the fact that the shifting sand and silt would build up and alter the course of this channel. Echo soundings and aerial photographs revealed the mass of metal was lying at a minimum depth of 14 feet below low tide level with about 3 feet of jagged metal obtruding above the river bed. The contractor concluded that it would be wise to abandon thoughts of scrap recovery and that the metal should be dispersed by divers detonating explosives to settle these spans on the river bed in small sections below the level prescribed by the river authority. The size of the charges was carefully arranged to avoid damaging local fisheries or disturbing birds in the Slimbridge sanctuary. Meanwhile the crane had deposited most of the remaining 132 feet spans on the Sharpness side before it had to return to Germany on 16th September, 1967; the hire charge being £21,000. Work had been slower than anticipated and left three spans, 21 piers and the swing bridge still in place.

Following the departure of *Magnus II*, the Nordman Construction Company purchased, for £3,250, the *Severn King* formerly owned by Old Passage Ferry which had become redundant with the opening of the Severn road bridge in 1966. At Gloucester, replacing the vessel's turntable, a Ruston caterpillar-track crane, compressor and winch were installed in order to load the scrap metal for transport to Sharpness Docks.

The Western Region District Civil Engineer demanded an explanation from Nordman Construction why the viaduct carrying the bridge line over the South Wales line would not be demolished by 31st December, 1967 as expected.

Swinnerton & Miller Limited of Willenhall were sub-contracted for the task. Holes were drilled at the bases of the piers and later extra vertical holes were driven from the tops of the piers adjacent to the South Wales main line to ensure successful fragmentation.

A 16-hour possession of the main line was secured. Then at 3.00 am on 10th March, 1968, both the Up and Down main lines were lifted each side of the viaduct, over 1,000 pounds of explosives placed in the stonework and at 7.30 am workmen retreated to the safety of Severn Bridge station. At 7.38 three blasts were sounded on a Klaxon horn followed by a further blast a minute later; at 7.40 am clouds of smoke appeared along the viaduct followed seconds later by the thunder of high explosives and falling masonry. As the smoke cleared it could be seen that five arches on the eastern side of the main line were still standing, the charges having merely blown off the facing of coursed masonry leaving in each pier a core of exceptionally large stone blocks.

The debris on the site of the main line was cleared using tractor shovels and dozers, the track restored and eventually opened at about 10.00 pm. During the week those five arches were re-drilled and being situated away from the railway, were demolished on 15th March on a 'between trains occupation' basis.

In November 1968 the Nordman Company went into receivership and the contract terminated. BR decided to complete the work itself and appointed V. Harris of the district engineer's department as demolition supervisor. They arranged with Swinnerton & Miller Limited to carry out blasting the remaining piers and to purchase the *Severn King* and equipment.

On 4th July, 1969 at the evening tide, the bow mooring rope broke and the *Severn King* swung round and drifted on to the stump of Pier No. 2, creating a gash in its hull 2 feet long and 6 inches wide. With leap tides imminent, salvage was impossible. On 27th July a concrete box was placed over the hole. Two Uniflot pontoons towed from Sharpness were fixed to the sides and all bulkhead doors sealed before being refloated on the high evening tide of 28th July. Its final voyage ended in it being beached near the entrance to Sharpness Docks. The swing bridge was cut up in 1970.

It was a wise decision to demolish the bridge as it had been struck by shipping no less than seven times between 1939 and 1961. At half-tide the water rushed past the pier in 10-knot mill-race and if a ship had collided with the bridge at this speed just as a passenger train was crossing, the result would have been unthinkable. Piers Nos. 18, 19 and 20 by the main channel were protected by small fenders, but the cost of providing substantial protection would have been prohibitive.

Except for traffic to and from the docks, Sharpness closed to goods on 3rd January, 1966, much of the remaining traffic connected with the nuclear power stations at Berkeley and Oldbury.

31st March, 1999 – 1st April, 1999 the Royal Train was stabled overnight on the Sharpness branch, top and tailed by No. 47792 *Saint Cuthbert* and No. 47798 *Prince William*. They worked the 08.45 Berkeley Road Junction – Gloucester where No. 47992 was uncoupled and No. 47798 alone worked the 09.30 to Bristol Temple Meads for Her Majesty to distribute the Maundy Money at the Cathedral.

No. 47792 *Saint Cuthbert* in RES livery with No. 47798 *Prince William* at the rear, approach Berkeley Road Junction with the Royal Train to form the 08.45 working to Gloucester where No. 47792 would be detached. No. 47798 then worked the 09.30 Gloucester-Bristol Temple Meads conveying the Queen and Royal Party to the Maundy service at Bristol Cathedral, 1st April, 1999.
Richard Giles

Chapter Seven

Description of The Midland Railway's Branch from Berkeley Road to Lydney Town

The Midland Railway (Additional Powers) Act of 25th July, 1872) authorised that company to lay a branch 4 miles in length from its Gloucester to Bristol main line at Berkeley Road, through Berkeley to the docks at Sharpness. It was an easy line to make as few earthworks were required and it opened for goods traffic on 2nd August, 1875 and following the installation of the block telegraph, to passengers on 1th August, 1876. Interestingly the local press seems to have ignored both these events.

The GWR had running powers over the MR line from Gloucester to Bristol and when, in 1894, the branch was absorbed into the Severn Bridge line, it gained powers to run from Berkeley Road to Sharpness, this providing a useful diversionary route for use when the Severn Tunnel was closed for maintenance. The fact that the junction at Berkeley Road faced north meant that a reversal was required. To avoid delays caused by this and offer a direct run, on 9th March, 1908 the GWR opened a double track loop from Berkeley Road South Junction to Berkeley Road Loop Junction. In order to deal with construction trains, the Loop Junction signal box was opened 29th November, 1904. The boxes belonged to the MR, but the loop itself was GWR property.

The MR, and later the LMS, ran a daily goods train from Sharpness to Bristol, but GWR goods trains over the loop were far more numerous. It is believed that the last train over the loop ran in November 1960 and it was closed on 27th January, 1963, its main use as a diversionary route being made redundant by the collapse of the Severn Bridge.

As an economy measure, the line from Berkeley Road to Sharpness South Junction had been singled on 26th July, 1931. A length at Berkeley Road was retained to provide a long refuge siding for Up main line freight trains, but

The Railway Correspondence & Travel Society special approaches Berkeley Road Loop Junction on 26th September, 1959. It is signalled to take the branch to Sharpness.

Dr A. J. G. Dickens

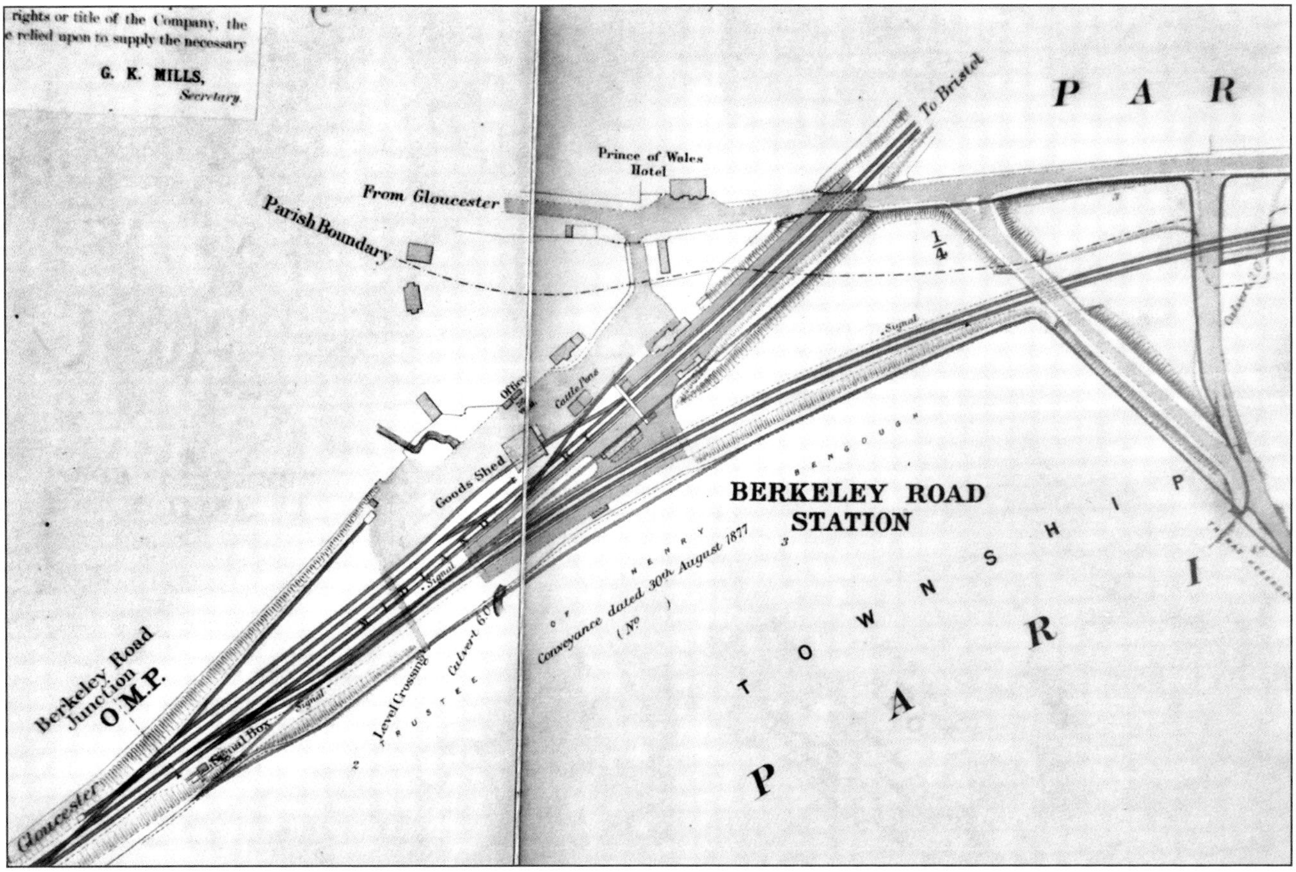

Berkeley Road plan 1898. The line from Gloucester and the north is in the bottom left corner. The line to Sharpness and the bridge leaves on the right-hand side.

latterly used for storing surplus wagons. Passenger traffic ceased on 2nd November, 1964, trains on the last day being worked by 0-4-2T No. 1453, but the line remained open for goods traffic from Sharpness and also nuclear power workings from the siding at Berkeley.

Description of the route

Berkeley Road station (108 miles 00 chains from Derby), initially 'Dursley & Berkeley' until 1845, opened on 8th July, 1844 to serve Dursley 3 miles distant and Berkeley 2½ miles away. Brunel being the Bristol & Gloucester Railway's engineer, the architecture was typical of his style.

The main building on the Down platform was of brick relieved with stone, and was given a bay window, (curiously with a chimney above), a horizontal canopy and tall chimneys. The Up platform had a brick and stone waiting shelter, but most of its horizontal canopy was removed early in the BR period. The three running-in boards proudly proclaimed 'Berkeley Road Junction for Sharpness and Severn and Wye Railway' until repainted by BR about 1950 when it was modified to 'Berkeley Road Change for Sharpness-Lydney'. In the 1950s the station garden was a delight. The goods yard, at the Up end of the line on the Down side, had a large brick goods shed. As the Bristol & Gloucester was originally broad gauge, this shed was suitably built to accommodate it. Adjacent were cattle pens and stables for the shunting and delivery horses.

The original tall signal box opened in 1870, was replaced on 27th July, 1900 by a standard MR box which was reduced to a ground frame 14th October, 1968. This in turn was taken out of use 15th February, 1970 and replaced by another 4 chains to the north. This had a short life and was closed in June 1972, the junction then remotely controlled from the MAS box at Bristol.

Berkeley Road view Up, 28th August, 1948. The running-in board reads: 'Berkeley Road for Sharpness and Severn and Wye Railway'.
Author's collection

Running-in board 1964, Lydney covered over after the 1960 bridge collapse.
D. Payne

2-4-0T *Friar Tuck*, later MR No. 1124A, at Berkeley Road.
Dr Budden

Berkeley Road view Up 1883. The standard Midland Railway footbridge is the first of its type.

Author's collection

The opening of the Sharpness branch required two additional platforms (0 miles 12 chains from the junction) to serve branch trains. A 59-feet span footbridge was added in 1883, its landing supported by cast-iron classical columns. Passengers from the Up branch platform had no such safety luxury and proceeded over a wooden crossing. A timber Midland-design building with fretted valance and hipped roof was built at the north end of the combined Up main line platform and the Down branch platform. It had no less than three fireplaces. It is rumoured that the shelter on the Up branch platform was once blown into an adjacent field, was recovered and re-set.

Latterly only the Up platform was used by branch passenger trains, the edging to the branch Down platform being removed. The station closed to branch passengers 2nd November, 1964, and to main line passengers 4th January, 1965; goods traffic, latterly mostly house coal, ceased 1st November, 1966. All buildings were soon demolished except for the goods shed which lasted another 11 years, so today only the station master's house remains, still in residential use.

South of the station the A38 crossed the main line at an oblique angle by a brick bridge. Around 1973, when the road was widened, a steel girder span was added, later to be replaced by a concrete arch. As it was originally an MR line, it is hedged rather than wire fenced.

The junction between the main line and the Sharpness branch was singled 2nd May, 1965 when the connection between the Down main line and the Down branch was taken out of use.

Leaving Berkeley Road station, initially level, at 40 chains it falls at 1 in 200 passing Baker's Siding, (1 miles 16 chains) opened 1st April, 1906 and closed during the First World War, before arriving at Berkeley Loop Junction (1 mile 26 chains) where the loop from the main line at Berkeley Loop South Junction offered direct access to and from Bristol.

Opened by the GWR 9th March, 1908, it allowed through running to and from Cardiff, Bristol, Portsmouth and Plymouth when the Severn Tunnel was closed for maintenance. The Midland used it for a daily goods train from Sharpness to Bristol until 1945, but GWR goods trains were more numerous; the only passenger trains which used the line were Sunday diversions. Berkeley Loop Junction closed 24th March, 1963. The loop, 1 mile 22 chains in

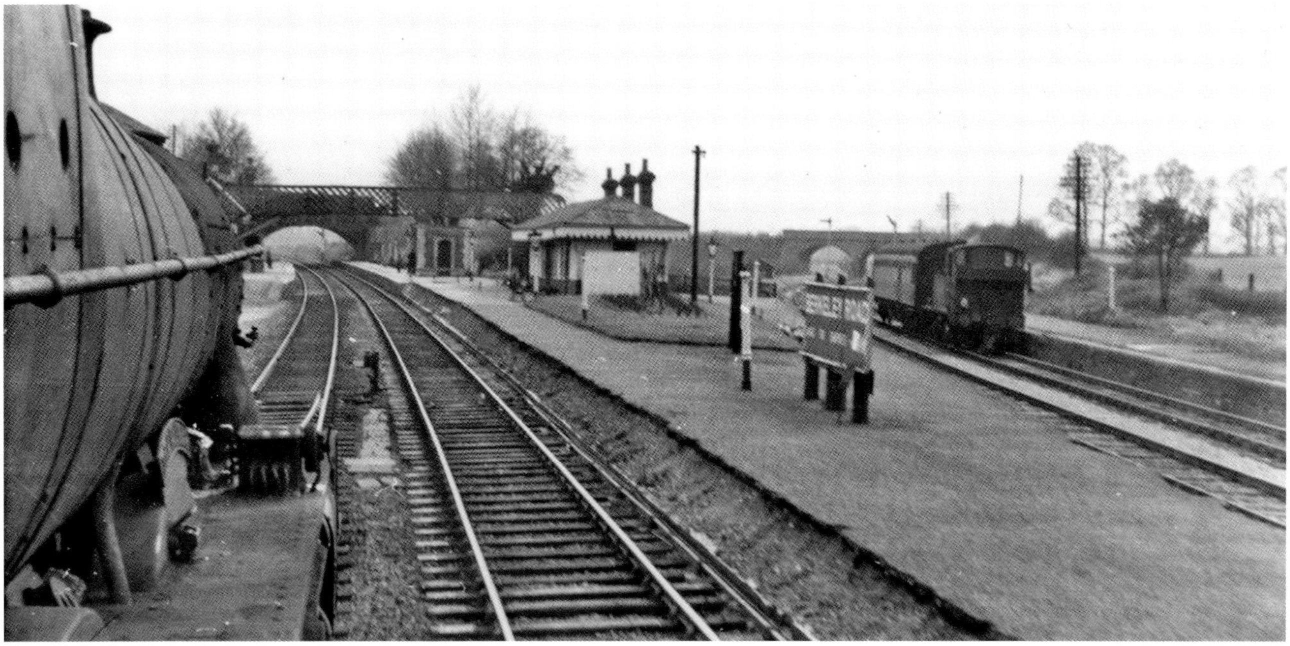

Top: View Down from Berkeley Road signal box, June 1914. On the right is a corrugated-iron lamp hut.
Author's collection

Centre: A 'Jubilee' class 4-6-0 heading a Down train passes Berkeley Road April 1964. Notice the cut-back platforms for shorter trains and the Sharpness train consisting of an auto car and a 0-4-2T.
W. F. Grainger

Right: Berkeley Road view Up 1949.
Author's collection

View Up from footbridge of Berkley Road, 1964, showing the goods shed. *D. Payne*

The 1.10 pm Saturdays-only Bristol-Gloucester headed by Class '4F' 0-6-0 No. 44272 enters Berkeley Road, 14th June, 1958. Unusually the Lydney train on the right is being hauled, not propelled. *R. E. Toop*

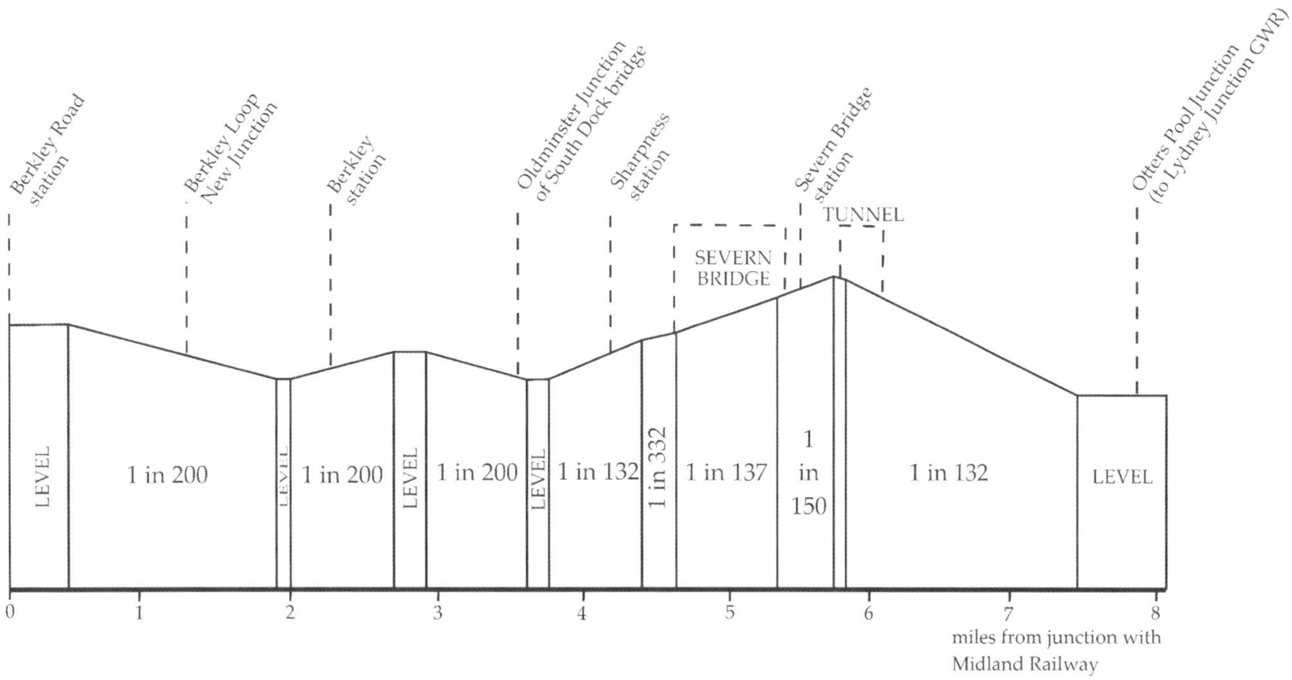

Gradient profiles: Berkeley Road Loop – Otters Pool Junction

length, was on a generally falling gradient from the main line. The loop was subject to a limit of 15 mph throughout. The signal box was only normally switched in from 7.00 pm until the small hours, though it was open for a longer period on winter Sundays when it was used as a diversionary route.

From Milepost 2 the Sharpness branch rose at 1 in 200 to Berkeley station (2 miles 13 chains). Set ¾ mile from the town, it had a typical MR style twin-pavilion hipped slate roofed building, of two colours of brick relieved with stone, on its Down platform. A prominent feature was the 100 or so attractively-shaped soffit brackets supporting the roof. The centre portion giving access to the platforms via an enclosed porch area, was fronted by an attractive glazed and cast-iron framed screen with a skylight above. The Up platform offered passengers a wooden shelter. The station was designed by John Holloway Sanders, the MR's chief architect. The passenger station building, goods shed and station master's house were

Berkeley, view Down, showing the details of the station building.
Lens of Sutton

Staff outside Berkeley signal box *circa* 1920; closed when the line was singled 26th July, 1931, it was one of the very few Midland Railway signal boxes with a brick base. The 'S' requests the lineman to call.

Peter White collection

Albert Baker (left) and Elton Vivian King at Berkeley station in the early 1920s. Their caps are lettered 'GW & MID'.

Peter White collection

constructed by John Roach of Charfield. In later years, as passenger trains did not exceed two coaches, the west end of the platform was fenced off. Sadly the station was demolished in the late 1970s.

An early example of the standard MR signal box, but with a brick base, stood opposite the goods shed with a 30-hundredweight crane, but with rationalisation, the box closed 26th July, 1931, after which either the guard or porter worked the north or south ground frames unlocked by the electric train staff. The brick base of the signal box was transformed into a platelayers' hut. A tramway linking the Down platform with the Berkeley Vale Dairy, was modified in 1917 to run to the cattle dock but closed in 1931.

When the line was singled, the crossing loop held 60 wagons. The station closed to passengers 2nd November, 1964 and goods 1st November, 1966 and most of the sidings taken out of use 8th June, 1968. The Vale of Berkeley Railway Trust has excavated the station foundations and hopes to rebuild the station and also preserve the weighbridge hut. The goods shed was demolished *circa* 1964 but the station master's house is still extant.

Closure in 1966 was not the end of the siding at Berkeley, as an overhead gantry crane had been installed for the transfer of nuclear waste flasks from two power stations at Berkeley and Oldbury. That at Berkeley opened in 1962; its Reactor No. 2 shut down in October 1988 and No. 1 in March 1999. Oldbury opened in 1967 and was decommissioned in 2012. From March 1999 all flask trains on the branch were operated by Direct Rail Services, a subsidiary of Nuclear Decommissioning Authority using refurbished 'Class 20' first-generation diesel-electric locomotives; currently they are worked by 'Class 88' diesel-electric locomotives run on an almost weekly basis. As there is no run-round at Berkeley, trains travel first to Sharpness where the engine can run round before returning 1½ miles to the trailing siding at Berkeley. The driver of the weekly nuclear flask train has to collect the single line staff from the signalman at Cheltenham – he's the nearest.

Berkeley view Up *circa* 1910 in double track days. A portion of the waiting shelter is on the far left with the signal box, closed 26th July, 1931 and the goods shed in the centre. Quite prominent is the loading gauge.

Author's collection

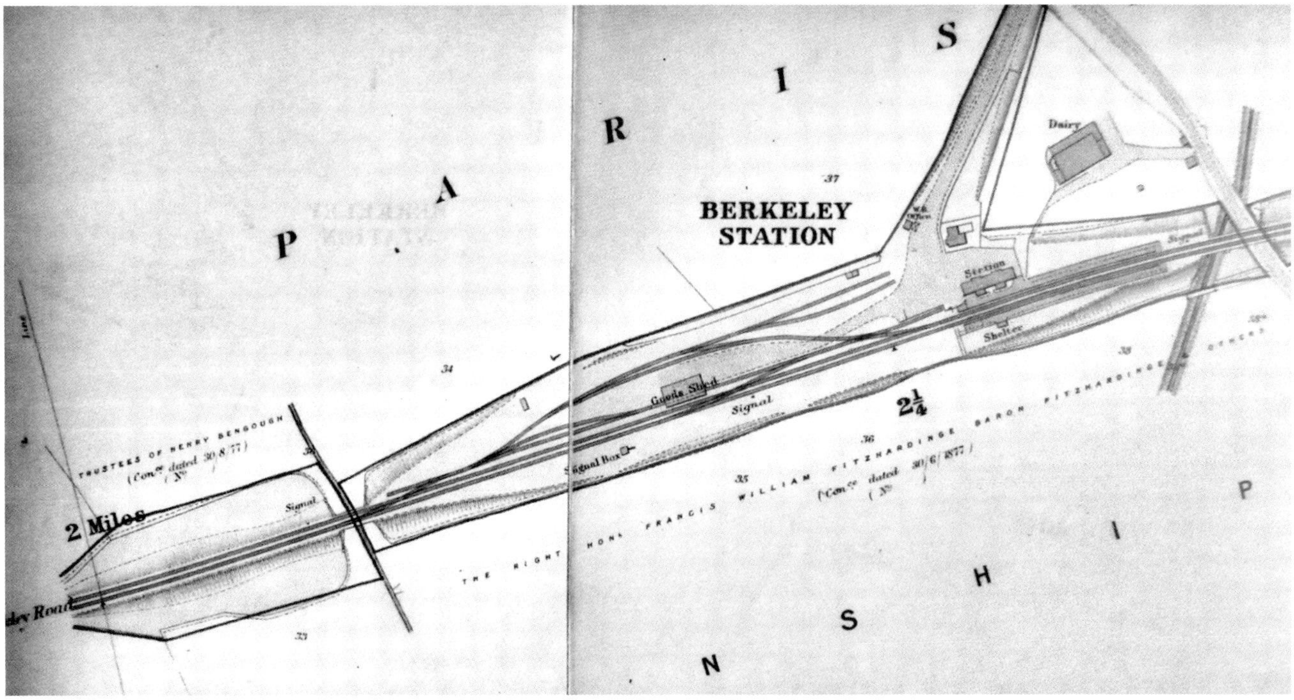

Berkeley plan 1898.

A Down propelled auto train approaches Berkeley. The coach is a Collett non-corridor brake-third adapted for push-pull working.

Author's collection

Berkeley, view Up, September 1964. Note the gantry crane for handling nuclear material.

D. Payne

The final EWS (BR)-operated nuclear flask train to run prior to the full DRS (Sellafield-based) takeover. On 25th February, 1999 No. 37375 in the Central Electricity Generating Board Berkeley compound displaying the special headboard reading: '7M56 1322 Berkeley to Sellafield last E.W.S. operated flask train 25th February, 1999'. The train was operated that day with a top-and-tail formation by three Bescot-based drivers; outwards via Bristol Parkway due to a points failure at Berkeley Road Junction, but returned double-headed with No. 37689.

Richard Giles.

Berkeley: entrance to the nuclear flask siding, 10th April, 1991.

Author

Berkeley, 10th April, 1991: remains of the Down platform centre right and the nuclear flask gantry in the background.

Author

Type 1 diesel-hydraulic 0-6-0 D9522 working the 13.20 Sharpness–Berkeley Road climbing the 1 in 200 out of Sharpness, 1st June, 1967.

Author

The first Midland Railway passenger trains from Berkeley Road terminated at a temporary single track timber-built station at Sharpness (3 miles 50 chains). Immediately before the passenger station was a signal box which has enjoyed many name changes. Originally Station, it became Docks Junction in 1879, (it formed a junction with the South Dock Branch opened in 1875), then Oldminster Junction and later still South Junction. The crossing loop at South Junction could hold 60 wagons. On 25th January, 1914, it was replaced by a new box of that name positioned in the 'V' of the junction. As at Berkeley, the brick base of the box was roofed to turn it into a permanent way hut.

In 1898 the MR installed a 45-feet 9-inch diameter manual turntable for the 0-6-0s operating daily freight services from Bath, Bristol and Gloucester; it was removed in the late 1950s but its brickwork can still be seen.

The second Oldminster siding constructed *circa* 1881 and now the current Road 1, had, and still has, a standard MR sleeper stop block, once very common, but now rare. The design was cheap both to construct and maintain. It consisted of sleepers measuring 84 inches by 11 inches by 5 inches buried 2 feet in the ground to form a stockade 16 feet long and 7 feet wide. At its face, the sleepers were set edgeways to give strength.

The structure was strapped with double-headed rail rails to prevent the ash, when compacted by vehicles striking it, splitting the block's sides. Most versions had a rear sleeper wall, but that at Oldminster is open-ended. It could stop low speed collisions with relatively minor consequences, as the sleepers and ash compressed on contact. If speed was higher, vehicles usually reared up, thus preventing forward movement and absorbing the shock more slowly.

This 900 feet long siding was used by the MR as an exchange siding where wagons of imported commodities, such as sawn timber, would be placed by the dock shunter to be collected daily by the pick-up goods.

A second siding with a rail-built stop block, was also added about 1890 and the other two added around 1940 also had this feature. These sidings fell out of use when the docks railway closed in 1991.

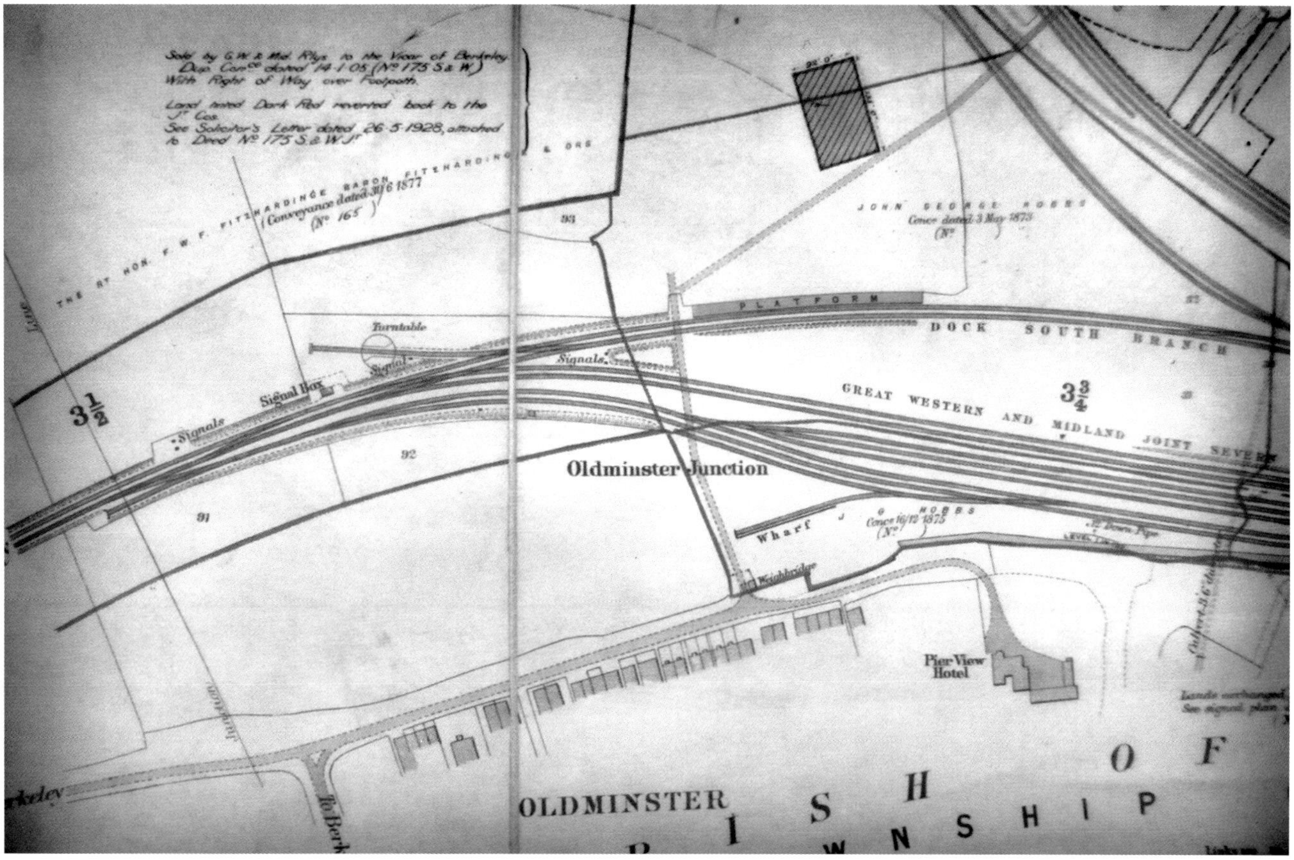

1898 plan showing site of original MR station at Sharpness.

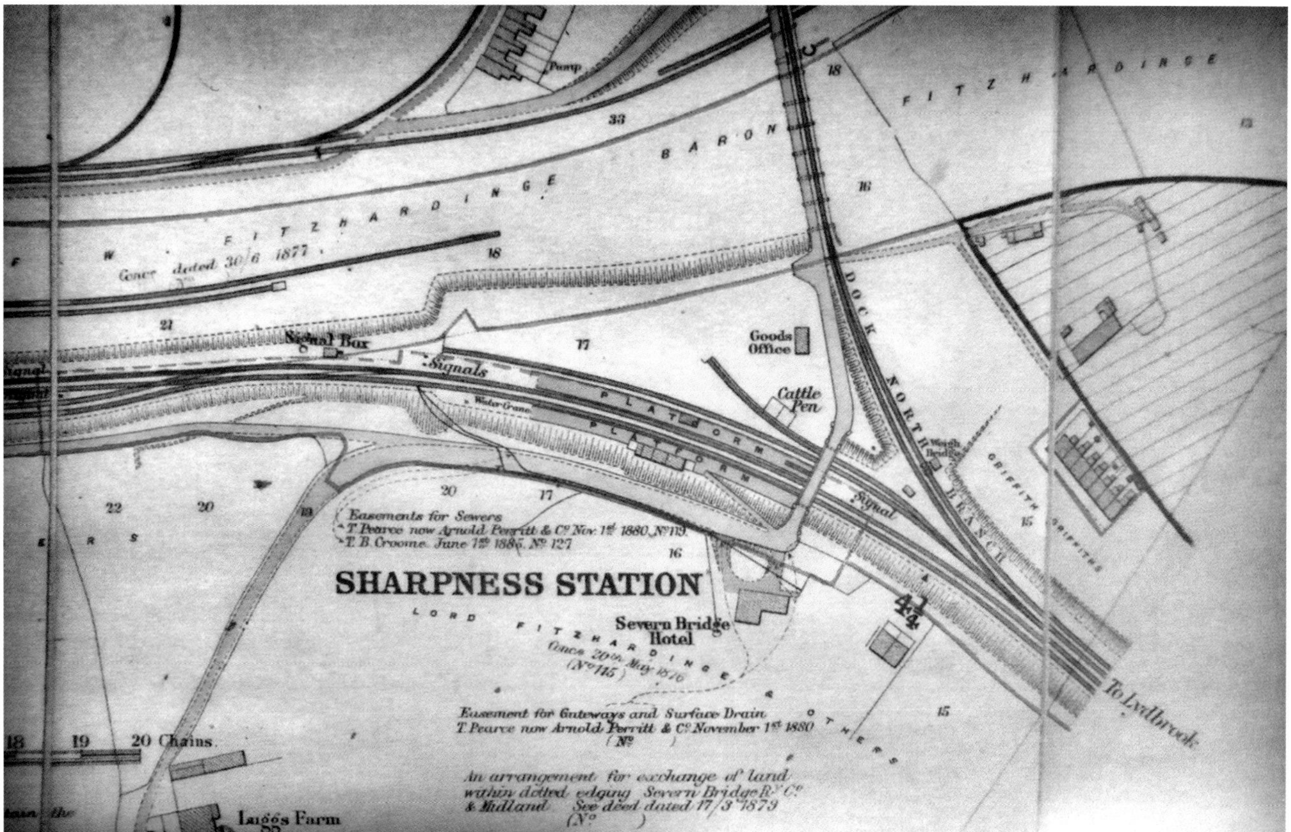

Sharpness station plan 1898.

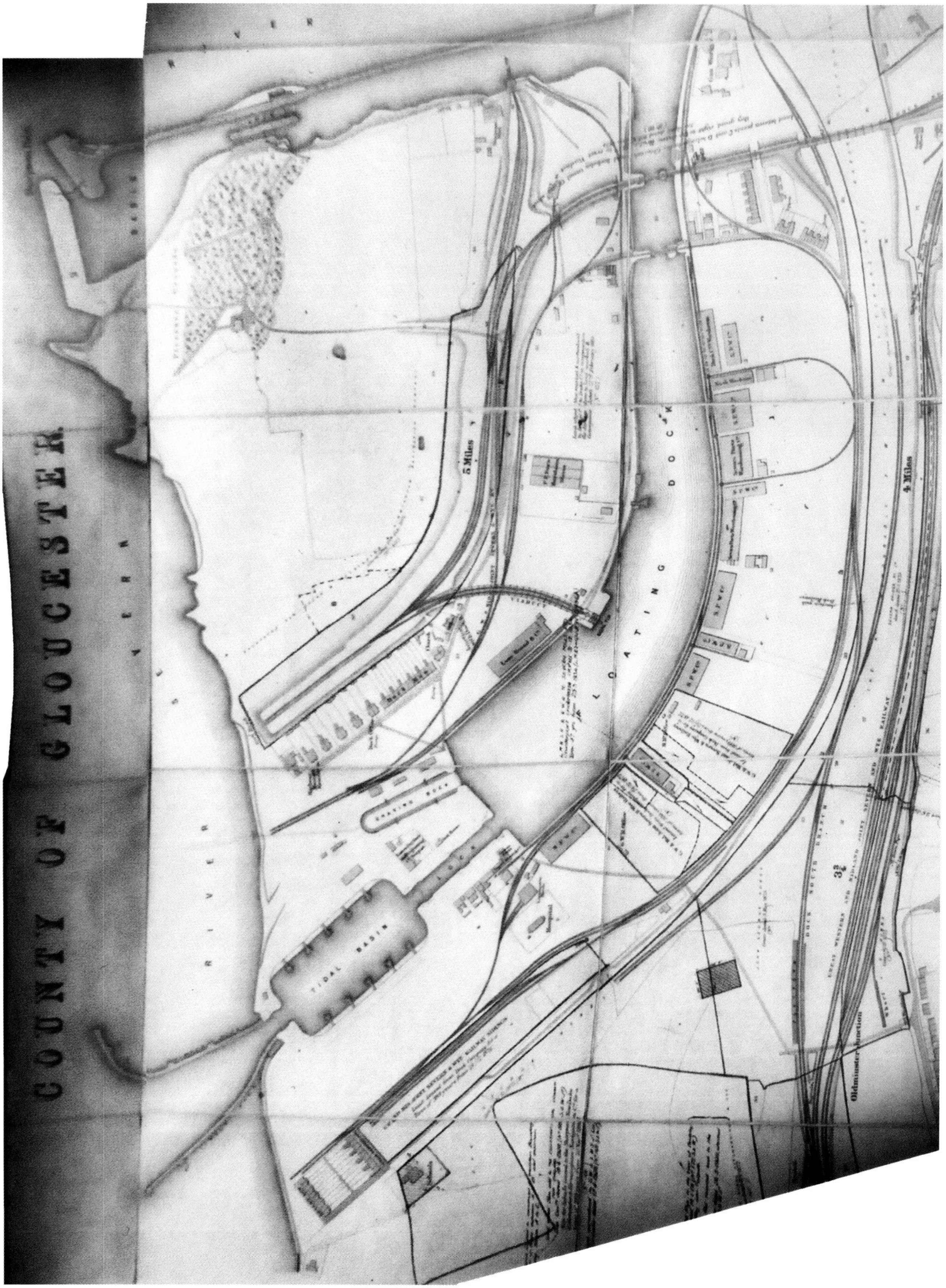

Sharpness Docks plan 1898, The platform at Oldminster is in the bottom right corner on the southern edge of the plan.

Oldminster footbridge about to be dismantled *circa* 1980.
John Mann

Nearby, Oldminster footbridge carried the Severn Way footpath linking Newtown with the docks. Built in 1905 it consisted of three spans: 48 feet 9¼ inches, 54 feet 7½ inches and 44 feet 9 inches. Dismantled in the late seventies or early eighties, it reverted to a foot-crossing but the brick abutments may still be seen. At Oldminster, or South Junction, the goods lines diverged east of the main line and the dock lines diverged west. In 1887 the Severn &Wye put in a wharf and two sidings for local goods traffic.

The Midland Railway passenger station at Sharpness closed on 16th October, 1879 and was replaced on 17th October, 1879 by a double track station on the Severn Bridge line at 4 miles 15 chains. Here the offices were in a brick structure with a hipped slate roof and a narrow, lead-covered canopy above the platform. Unusually, its rear wall was blank, the only access being via the platform. From 1903 a brick shelter on the Down platform formed a lean-to frontispiece to Sharpness Station signal box, this 33-lever box closing 27th October, 1957 when the responsibility of controlling Severn Bridge from the east was transferred to Sharpness South and key token working Berkeley Road – Sharpness replaced Tyer's system. The station garden was well-maintained. The loop at the station was removed in September 1956. Except for traffic to and from the docks, the station closed to goods 3rd January, 1966; until closure it was gas-lit.

As the original coal tip was situated on the relatively shallow Gloucester & Berkeley Canal, it had proved unsuitable for larger vessels and so a deep-water tip was constructed approximately midway along the western side of the floating dock. The new tip was first used in April 1886, but fell out of use in the post-Second World War years. Wagons reached the new tip along a substantial viaduct initially of timber, later replaced by concrete.

The original tip outlasted the latter, being used for Forest coal until *circa* 1965 before it was eventually demolished in 1972. Post-October 1960 Forest coal continued to arrive by rail but via Gloucester, before being tipped into barges for use at Cadbury's factory, Frampton-on-Severn. Access to both tips was via the high-level swing bridge. This raised line allowed loaded wagons

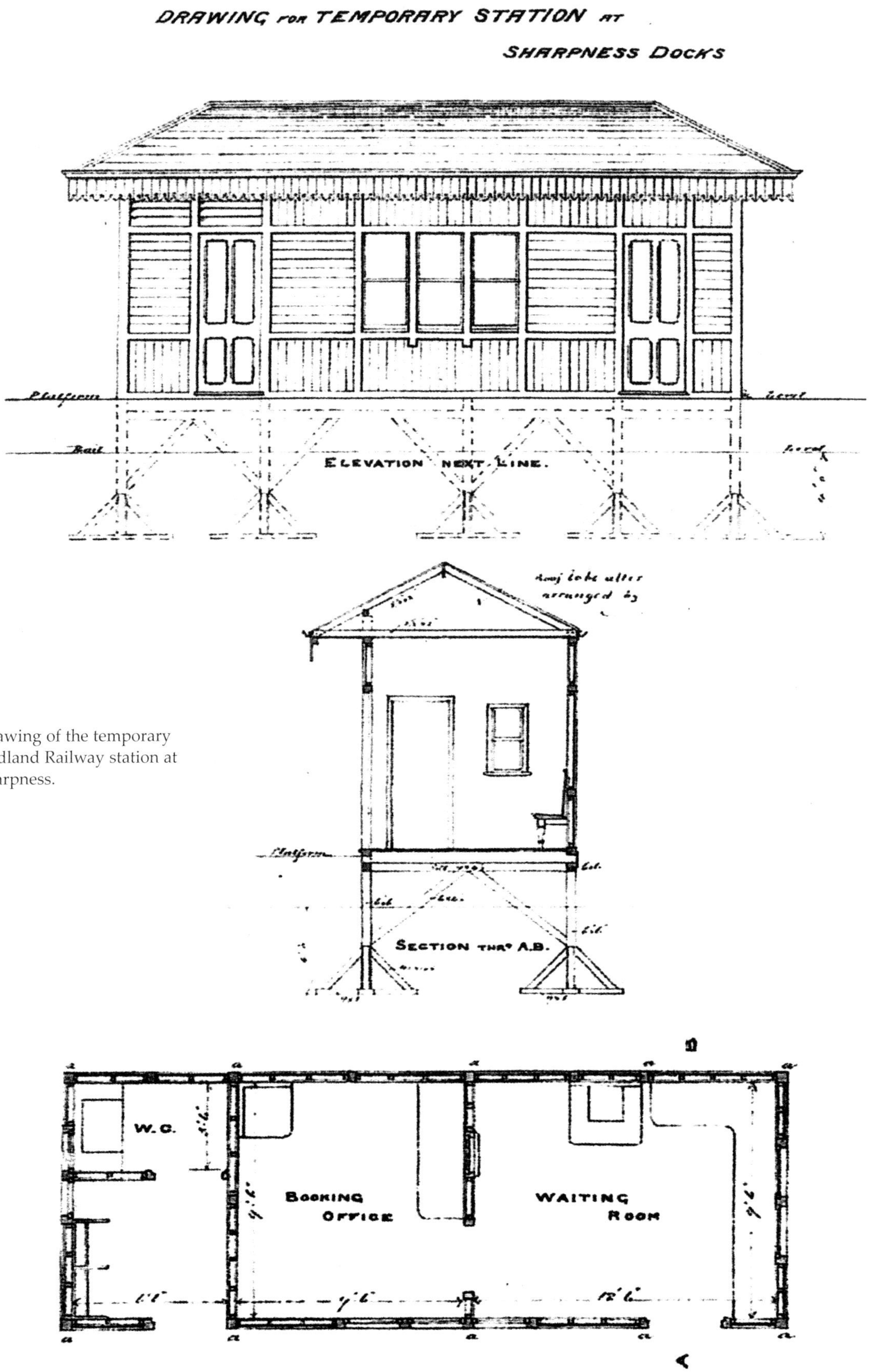

Drawing of the temporary Midland Railway station at Sharpness.

Sharpness Docks No. 5 0-4-0ST, an Avonside of 1924, outside its shed, 15th August, 1956. *Revd. Alan Newman*

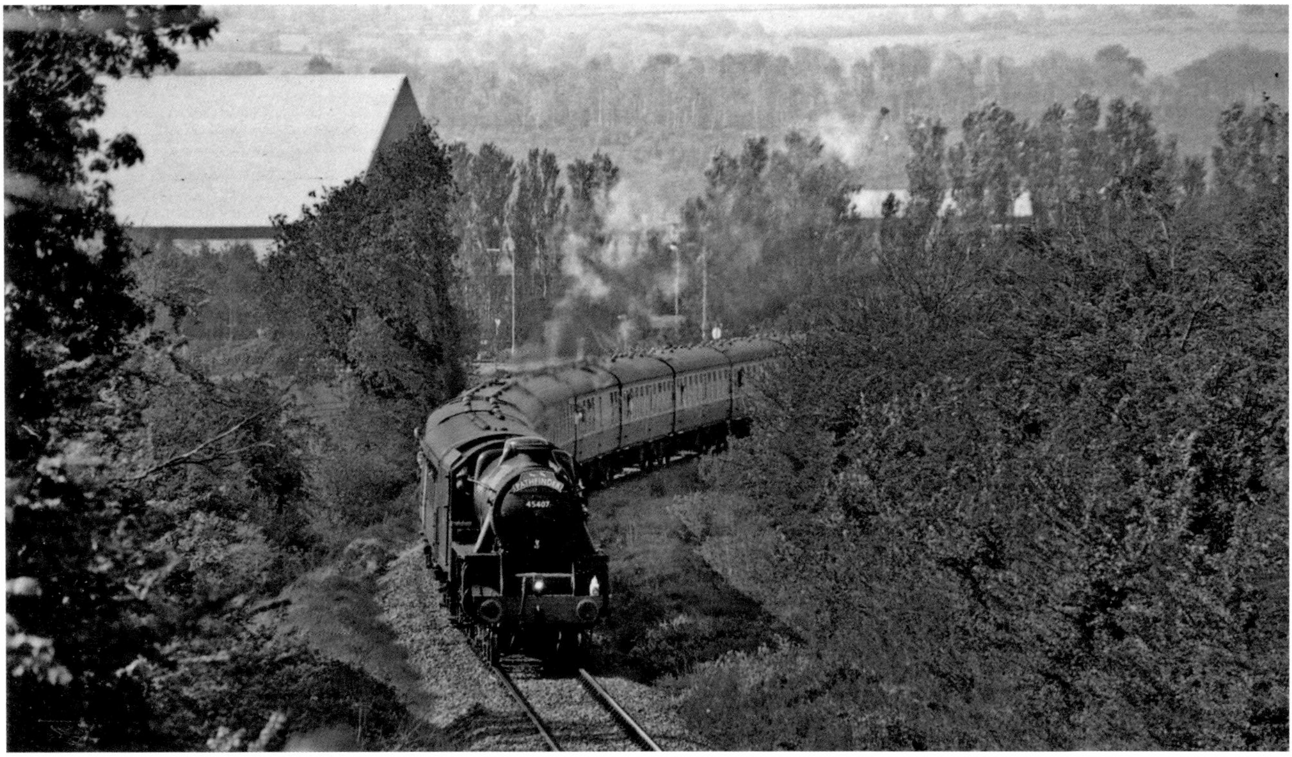

Ex-LMS Class '5' 4-6-0 No. 45407 *Lancashire Fusilier* works the Severnside Rambler on 22nd April, 2007 near Newtown; Sharpness Docks and warehousing are in the background. This was the first steam working over the freight-only branch for about 40 years.

Richard Giles

Wagon tip at Sharpness, 1st June, 1967.
Author

to reach a tip by gravity, and after unloading, to run away. The timber-built tip allowed for end-discharge wagons to lean to an angle of 45 degrees and shoot their contents into boats or lighters 35-40 feet below. Before reaching the tip, a turntable was provided to obviate the problem if the doors were at the wrong end.

In 1916 a factory at Frampton-on-Severn was opened to supply condensed milk to the army. Post-First World War Messrs Cadbury turned the factory over to making chocolate crumb. Cocoa beans arriving at Bristol Channel ports were unloaded into lighters, or barges and then taken up the Severn and the Gloucester & Berkeley Canal to Frampton, four miles north of Sharpness. Milk from nearby farms was combined with the prepared cocoa and sugar to make chocolate crumb which was then loaded straight from the factory into narrow boats, taken via the Severn and then the Staffs & Worcester Canal to the Cadbury Bournville factory, also on a canal bank.

Frampton used coal from the Princess Royal Colliery, Forest of Dean. It was taken over the bridge to Sharpness and loaded into barges using the 1880 coal staithe in the Old Docks. BR continued to use the old 7-plank wagons here long after most of the others of this pattern had been withdrawn, because they were the heaviest allowed on the counterbalanced tip. The use of the stage finally ended on 9th February, 1965 when the coal boilers at Frampton were replaced by those burning oil. The Old Docks Coal Staithe was demolished in August 1972 Cadbury's factory at Frampton closed in 1982.

Until the 1960s most of the port of Sharpness's activity was centred on grain and timber, much of it transhipped to barges or rail for Gloucester or the West Midlands.

Until *circa* 1955 when the gas works closed, an extension of the quayside line ran to the works situated beyond a high-level bridge on the south side of the Gloucester & Berkeley Canal.

Lighters by the timber ground; a line of flat wagons, left, 15th April, 1968. *Author*

Sharpness Docks *circa* 1910. The 'Gloster Steam Packet Wharf' is on the centre right with a passenger vessel moored alongside.
Author's collection

Upper: No. 2 swing bridge in the open position, 9th October, 1980. *John Mann*

Lower: No. 2 swing bridge with some remaining spans of the Severn Bridge visible beyond, 15th April, 1968. *Author*

Sharpness Swing Bridge East Ground Frame (No. 2 viaduct). *D. Payne*

The author's Ford Consul on the road/rail No. 2 viaduct swing bridge, 22nd August, 1956. The coal tip tower in the top right corner. *Author*

The low-level swing bridge (No. 4 viaduct) at Sharpness, 9th October, 1980.
John Mann

Signal protecting No. 2 swing bridge, 25th March, 1991.
Author

Type 1 diesel-hydraulic D9522 shunting at Sharpness, 1st June, 1967.
Author

The National Sea Training School based at Sharpness from 1939 until 1966, offered the line considerable passenger traffic as in that period 70,000 boys were trained, about 2,500 annually. To assist these cadets, pre-1960 working timetables indicated an unadvertised Mondays-only non-stop service Lydney Junction-Sharpness with the condition 'Runs for 10 or more passengers'.

The training vessel used was *Vindicatrix* moored at the present marina near the original entrance lock to the canal. An iron-hulled sailing ship built in 1883, she had an interesting history. Wrecked off South Africa in 1903, she was repaired and sold to a German shipping line in 1910, commandeered by the German Navy in the First World War and used as a submarine depot ship. Surrendered to the Royal Navy, she became a prison for Germans at Leith, before moving to the West India Docks as a seamen's hostel. She arrived at Sharpness on 1st September, 1939. Sadly, following the school's relocation to Gravesend on 12th January, 1967, she made her final voyage to a breaker's yard at Newport. At one time the musician Tommy Steele was a cadet aboard *Vindicatrix*.

The dock company's engine placed GWR and LMS traffic from the docks on the Joint Line sidings just outside the dock boundary fence and in a like manner, Joint Line engines placed all traffic and empty wagons on the dock company's sidings inside the boundary fence. In the case of LMS traffic and empties for the dock, the dock company's engine and men could fetch the same from the Joint sidings under the direction of the Joint foreman or shunter, but were prohibited from passing on, over, or across passenger running lines. A Stop Board showed the boundary and neither the Joint nor dock engines could pass the board unless accompanied by and in charge of the foreman or shunter on whose sidings they were about to pass.

Tenders in scrapyard, 20th July, 1964.
Richard Brown

Scrapyard at Sharpness, 20th July, 1964.
Richard Brown

From about 1882 when the dock company took over the North Dock Branch, opened in 1879, it owned a stud of locomotives and internal-user wagons. The first diesel locomotive arrived in 1961 and the second in 1963, the last steam locomotive then being placed in storage. Latterly the two diesels were occasionally used to shunt former BR china clay wagons, which had become redundant due to their replacement by air-braked wagons, arriving for scrapping together with the occasional wagons of railway scrap – mainly rails and chairs. Rail traffic ceased about 1989.

The internal-user wagons wore a livery of lead grey with white lettering and black strapping, solebars and buffer beams. Certainly some of these wagons were ex-Midland Railway, while in the 1930s about 50 open wagons were purchased from the GWR. Most of these internal-user wagons were scrapped about 1967.

The dock steam locomotives were kept in a two-road shed on the east side and at the south end of the floating dock. Latterly the engines carried the British Waterways roundel on their water tanks. The green-liveried diesel engines had a new one-road shed near the dry dock.

Principal commodities dealt with at Sharpness were grain and timber, and from the 1920s, also petroleum; the heaviest traffic was inwards. From the early 1960s the despatch of scrap metal became important. After the Second World War Alfred Cooper of Swindon set up 26 scrap depots, one opening at California Sidings, Sharpness. Between 1964 and 1965 25 locomotives were scrapped, all ex-GWR, including two 'County' class 4-6-0s No. 1006 *County of Cornwall* and No. 1027 *County of Stafford*. Until the early 1980s steel mineral wagons were also cut up and up to the early 1990s scrap metal for export arrived by rail before being shipped.

A new grain silo of 6,000 tons capacity was built in the 1970s.

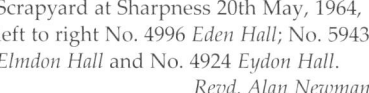

Scrapyard at Sharpness 20th May, 1964, left to right No. 4996 *Eden Hall*; No. 5943 *Elmdon Hall* and No. 4924 *Eydon Hall*.
Revd. Alan Newman

The passenger station at Sharpness *circa* 1910 with the signal box on the far right. The line to No. 1 viaduct on the left. The goods office is set between the viaduct and the signal box. The Severn Bridge is in the distance.

Author's collection

View from the docks past the goods office across the passenger station to the Severn Bridge & Railway Hotel, *circa* 1910.

Author's collection

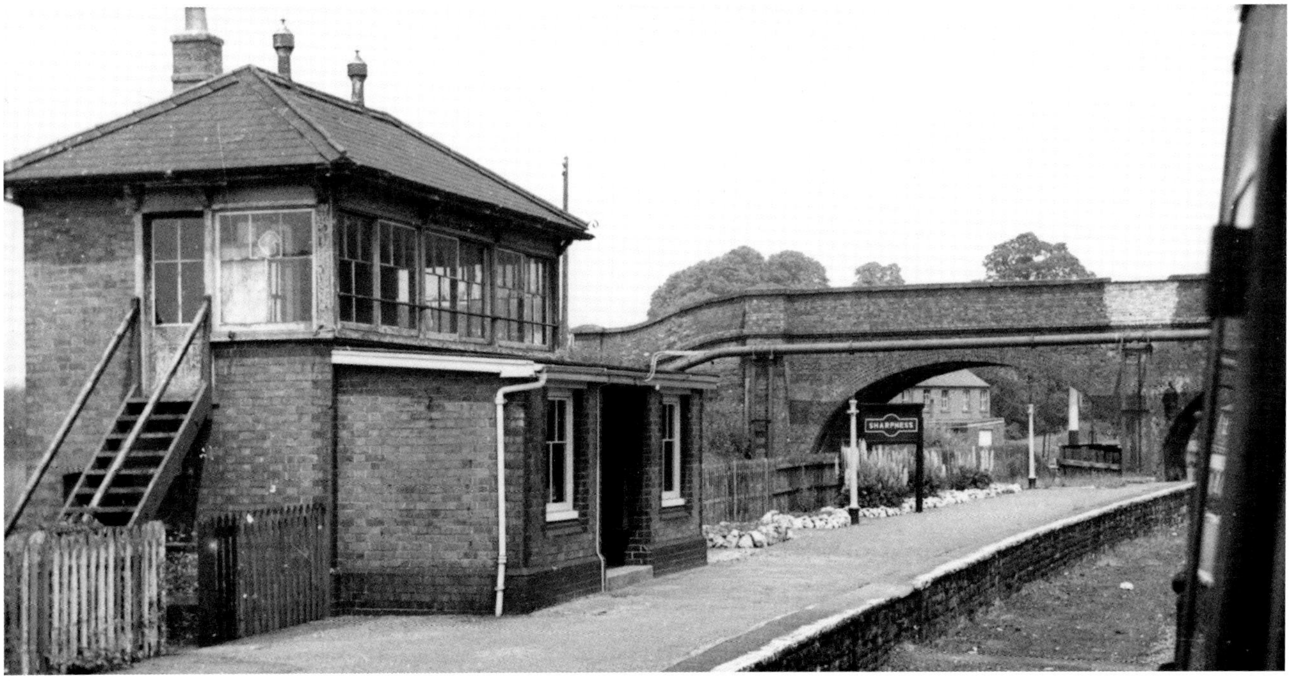

The unusual combined signal box and waiting shelter on the Down platform, 14th June, 1958. Notice the lifted Down track and the LMS running-in board.

R. E. Toop

'14XX' class 0-4-2T No. 1445 at Sharpness 5th August, 1964.
W. Potter

The GWR working timetable appendix for 1931 allowed a train from Lydney to run to the North Branch at Sharpness without going into Sharpness passenger station, the guard being held responsible for collecting the tablet from the driver and taking it to the signalman at Sharpness after seeing the last vehicle of his train clear of the main line.

Beyond the passenger station the line became single and on a rising gradient of 1 in 132, sharply curved through a cutting to cross the Severn Bridge. Above the 196 feet-long swing span at 4 miles 53 chains was Sharpness Swing Bridge signal box with seven levers, only three of which controlled signals, and one of these was for ships on the canal.

The swing span, weighing approximately 400 tons, crossed the Gloucester & Berkeley Canal. It swung on a table of rollers known as a 'live ring' set on a central masonry pier built between the canal and the Severn and was turned by a steam engine fixed in a glass-sided structure on top of the girders above the railway and over the central pier. This building also contained signalling equipment. To guard against breakdown, there were two sets of boilers, engines and machinery. Access to the cabin was via an internal stairway within the masonry pier.

Friction gear transmitted power from the engine to the machinery or locking gear. The locking and turning mechanism was designed so that both could not be in gear at the same time. The bridge was locked by iron wedges lifting it on the masonry piers. Moving simultaneously with the wedges was an iron bar sharply pointed at the end which, as it advanced, entered a socket in the pier and trimmed the rails to a nicety. As the bar was used to sever the telegraph, a train could not be accepted unless the bridge was locked. Later, electric interlocking prevented a token being withdrawn at Severn Bridge, or Sharpness boxes if the bridge was open. It opened turning in anti-clockwise direction. To prevent a derailed wagon damaging the bridge, a re-railer was fixed to the track and for safety a check rail laid over the bridge.

The eastern end of the Severn Bridge: notice the trap points, check rails and MR signal.

BR WR

Either an engine driver or signalman was required to be on duty at the bridge during the time the railway was open for traffic. The engine driver, (on the Lydney Locomotive Department payroll), had to be competent to work the electrical instruments, signals, locking apparatus and so forth and was required to do so when a signalman was not on duty. Likewise the signalman had to be competent to work the engines, turn the bridge and attend to the fires when an engine driver was not on duty. The latter was required to keep the engines and all the machinery clean and in good order, the signalman assisting the engine driver in coaling and cleaning the engines. Each engine had two cylinders of 8 inch diameter bore with a 16¼ inch stroke. They worked at a pressure of 60 pounds per square inch. The boiler also supplied power to the steam pumps providing water for a reservoir which fed the water cranes at Sharpness.

The job of engine driver was no sinecure, certainly not a matter of sitting back for most of the time and enjoying the delights of the Severn Estuary for he was required to:

1 Work a boiler for a fortnight before changing over to the other when it was to be thoroughly washed out, the firebars removed and tubes and firebox cleaned.
2 Test the safety valves daily by easing them.
3 Take out the lead plug in the firebox, examine and if necessary change it. The safety valves, connections, pressure gauges, blow-off cock and other fittings examined, cleaned at least every four months.
4 Blow the gauge cocks through at least once daily.
5 Check that at least two inches of water showed in the gauge glass.
6 Ensure that the working pressure of the boiler did not to exceed 60 pounds per square inch.

A form recording the above work had to be filled daily.

Additional requirements were:

a Only one set of turning and locking gear to be used but changed every four months.
b The locking indicator to be tested, and if necessary adjusted, every three days.
c The ends of the swing bridge be examined about mid-day each day to see that the signal plungers and pawls were clear; if any adjustment was necessary it had to be reported to the signal lineman. Should the ends of the rails touch, or the space be too great, it had to be reported to the inspector, or ganger of the permanent way.
d Record to be kept of the number of hours worked by the reservoir donkey pump in each day and then sent at the end of the month to the clerk of the works.
e The surface of water in the reservoir was not to be lower than six inches below the overflow.

The swing bridge machinery was maintained by the Gloucester Horton Road Locomotive Depot, but the bridge structure by the Gloucester Bridge Department.

During the time when the engine driver, or signalman was on duty, the bridge was required to be kept closed across the canal, but when no one was in attendance, it was to be kept open so that vessels could pass. A signal arm fixed to the bridge indicated to shipping whether it could be opened or not: a MR horizontal semaphore signal arm and red light showed that it could not be opened due to a train being in the section, whereas when the arm was lowered showing a green light, it showed that the section was clear of and could be opened. A vessel requiring the span to be swung sounded three whistles when it was at least 400 yards distant and when the span was fully swung the bridge operator replied with three whistles. After giving these whistles, vessels had to move slowly towards the bridge and be prepared to stop 100 yards from it in the event of it not being opened. Signalmen were required to watch for tall sailing vessels approaching and unable to give an audible warning.

British Waterways notice beside the Gloucester & Berkeley Canal with the Severn Bridge swing span open in the distance.

D. Payne

The Severn Bridge swing span and signal box 27th July, 1956, the doorway to the box at the foot of the stone column. BR WR

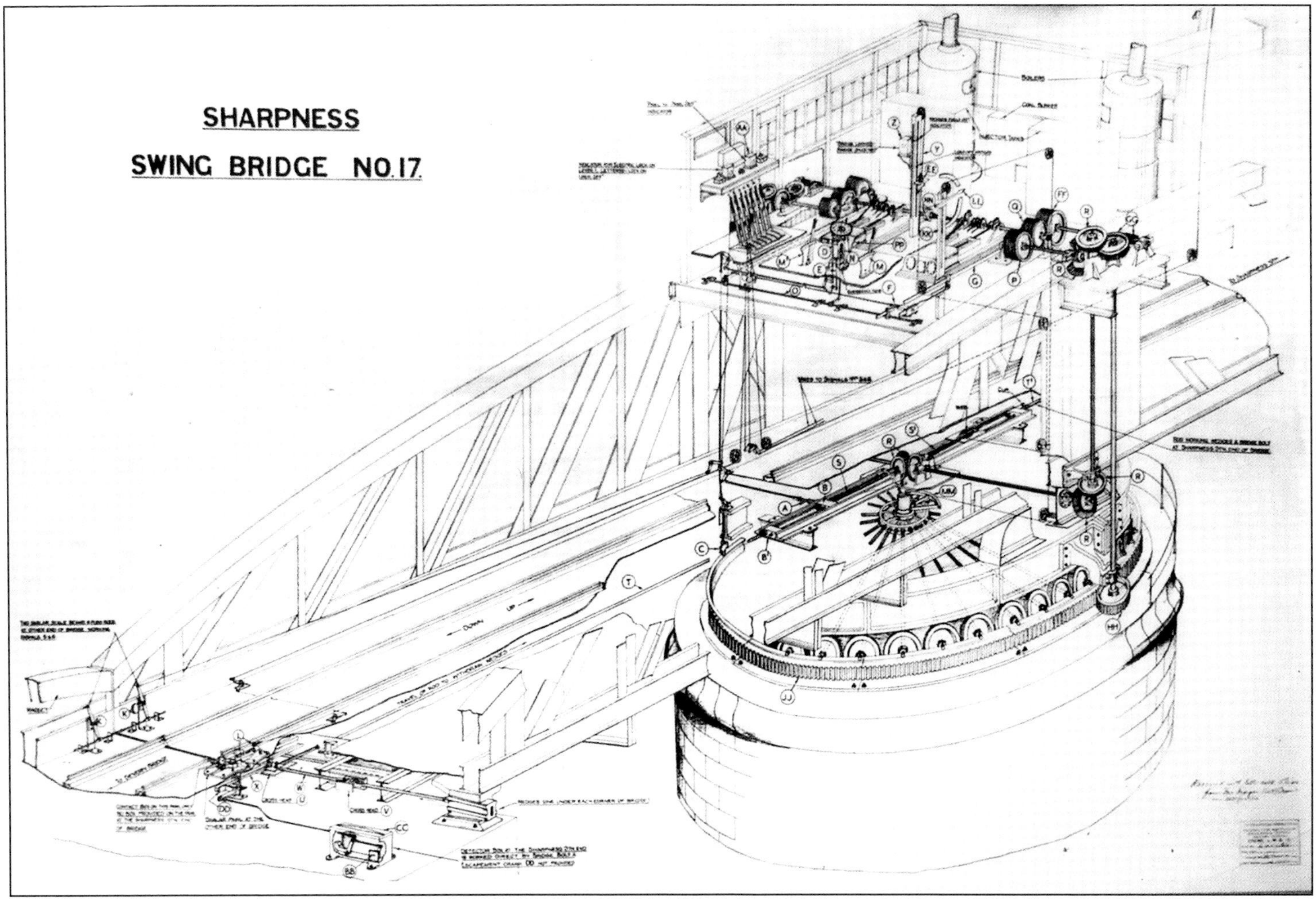

The Severn Bridge swing span. *BR WR*

The drive of the Severn Bridge swing span, 27th July, 1956. *BR WR*

The Severn Bridge swing span, 26th April, 1949, view west. Note check rails, the rollers below the span's centre, the MR-type signal and smoke from one of the boilers.

BR WR

A steam tug appears to be hauling several vessels through the open swing span *circa* 1910.

Author's collection

The Severn Bridge from the swing bridge signal box; note that the swing bridge was wide enough to allow for doubling; gas pipe on right.

BR WR

Strain gauges and wires on the swing bridge, 15th July, 1956.
BR WR

When the arm or light showed danger, vessels requiring the bridge to be opened had to stop at a white post on the canal towing path 300 yards from the bridge.

At night a white light was fixed at each end of the swing span to indicate to vessels on the canal when it was fully open parallel with the canal. Vessels navigating the canal in the dark were required to carry a white light at the bows and the signalman should not be requested to open the bridge for any vessel whose mast, or masts, could be lowered to pass underneath it.

To avoid danger to men at work painting, or carrying out repairs to the swing span, to prevent it being moved, three keys labelled 'Severn Bridge' were provided in connection with a lock on the clutch gear wheel. An endless chain apparatus was provided for conveying the keys between the signal box and the bridge floor. When a key was lowered, it was to be removed immediately from the chain by the gear cleaner, or other person in charge of the work. Should a train or vessel need to pass while work on the span was being carried out, the signalman gave three beats on the gong fixed outside his box and as quickly as possible, the men in charge of the work to had make it safe for a train or vessel to pass.

The bridge consisted of two 312 feet spans, five of 174 feet and 14 of 134 feet, all on a rising gradient of 1 in 137, easing to 1 in 150 at its western end where it crossed the South Wales main line. There was a 70 feet headway at high water below the two longest spans. The wrought-iron in the girders weighed 3,528 tons and there was 2,171 tons of cast-iron in the pier cylinders which were filled with 4,321 cubic yards of lime concrete. 1,500 tons of cast-iron were buried in the sand and went down to rock 70 feet below high water level. The swing bridge, 196 feet in length weighed about 400 tons. Prior to its westerly end, the bridge crossed the Gloucester-Newport main line by a 12-arch Forest stone viaduct about 70 feet high, each elliptical arch of 52 feet span. A restriction of 15 mph was in force over the bridge.

The Severn Bridge station end of the bridge; note the strengthening girders and the two lifebelts on the left, 1956.
BR WR

Roller (rocker) bearings between the two 312 feet spans, 15th July, 1956.
BR WR

Top: '16XX' class 0-6-0PT No. 1639 crosses the bridge with the 11.52 am Berkeley Road – Lydney, 22nd August, 1956.
Author

Centre: A passenger's view of an Up train leaving Severn Bridge station on 19th July, 1958.
R. E. Toop

Severn Bridge station: view showing the MR box opened 12th November, 1911.
Author's collection

For many years the bridge wore a livery of black for the lower part of the cylinders, the upper part chocolate with cream girders, but in the late 1940s and early 1950s, at a rate of two spans a year it was repainted in metallic grey. Most of this work was done by BR painters on timber trestles slung from girders by ropes and pulleys.

The bridge was entirely in Gloucestershire as unusually, the Severn here did not form a county boundary.

From 1930 to 1936 an anemometer was set on the bridge by the Department of Scientific Research, and in 1954 the Ministry of Transport & Civil Aviation set up four anemometers.

Severn Bridge for Blakeney station (5 miles 40 chains) had two platforms set on a curve; the crossing loop could accept 26 wagons. As the timber-framed station was set on an embankment, for lightness the platforms were constructed of wood, as was the signal box. The station building was provided by the Gloucester Carriage & Wagon Company in flat-pack form and erected on site. The original tall signal box at the north end of the Up platform closed 12th November, 1911 when a standard 12-lever MR replacement opened at the opposite end. The solitary goods siding placed between the station and the bridge, was taken out of use 24th June, 1956. Trains left the station on a rising gradient of 1 in 150. 70 yards in advance of the Down home signal was a runaway catch point worked from the signal box. The station was originally to be called Purton, but this was thought to be misleading, so was changed to Severn Bridge.

At 5 miles 62 chains the line entered the 506-yard long brick-lined, double-width Severn Bridge Tunnel, initially level, but soon falling at 1 in 132. The tunnel keystone bore the date 1874. If a freight train required banking assistance between Lydney Junction and Severn Bridge, it was to be coupled in the rear.

An enigmatic paragraph on an official 1924 plan states that 'The surface was restored for landowner's use, but right of way over tunnel reserved for opening shaft repairs etc when required' – though actually there was no shaft, but this paragraph would have permitted one to be made if ever it was found necessary.

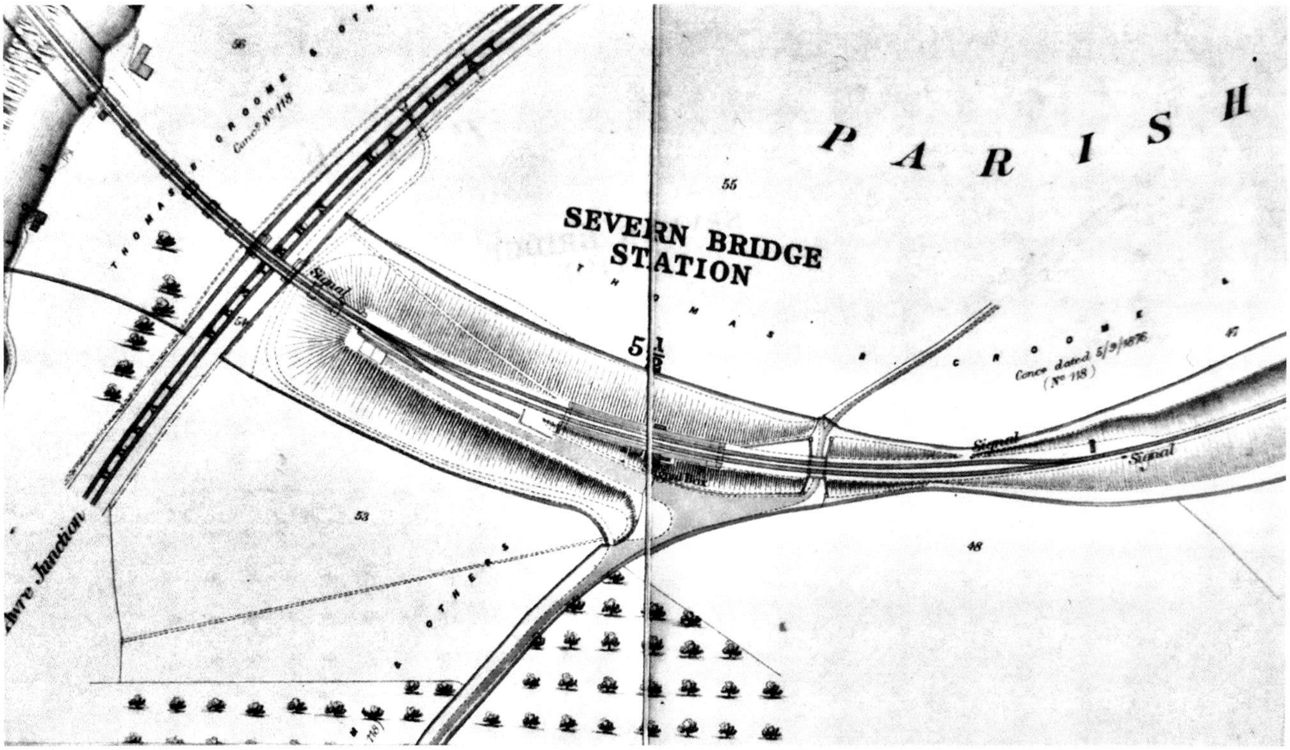

Severn Bridge station plan 1898. The stone viaduct leads to the bridge and Sharpness in the top left corner.

Top: Severn Bridge for Blakeney station 9th August, 1961. *Author*

Centre: Taken after the destruction of two of the spans, a gate across the entrance to the viaduct reads 'Danger'. A Hillman Imp stands on the cattle dock. *Richard Brown*

The north portal of Severn Bridge Tunnel. The notice on the right reads: 'Severn Bridge Tunnel 506 yards'; its lower left-hand corner has been repaired. This relic is now preserved by the Dean Forest Railway. The tunnel keystone is engraved 'AD 1878'. *A. T. Dowding*

View of the bridge from Severn Bridge and Blakeney station, 9th August, 1961. The horn for collecting the tablet pouch is on the right. Notice the Midland-type fencing. *Author*

The exchange of tablets at Severn Bridge, 19th July, 1958. The engine is 0-6-0PT No. 1639. *R. E. Toop*

Severn Bridge station, view Down *circa* 1965. The signal box windows have been boarded.

Lens of Sutton

A view after closure from just south of the station towards Severn Bridge Tunnel.

S. Apperley

'16XX' class 0-6-0PT No. 1664 climbs from Severn Bridge Tunnel with an Up Railway Enthusiasts' Club special comprised of brake vans. The occasion demands a brand-new bucket.

D. R. G. Nowell

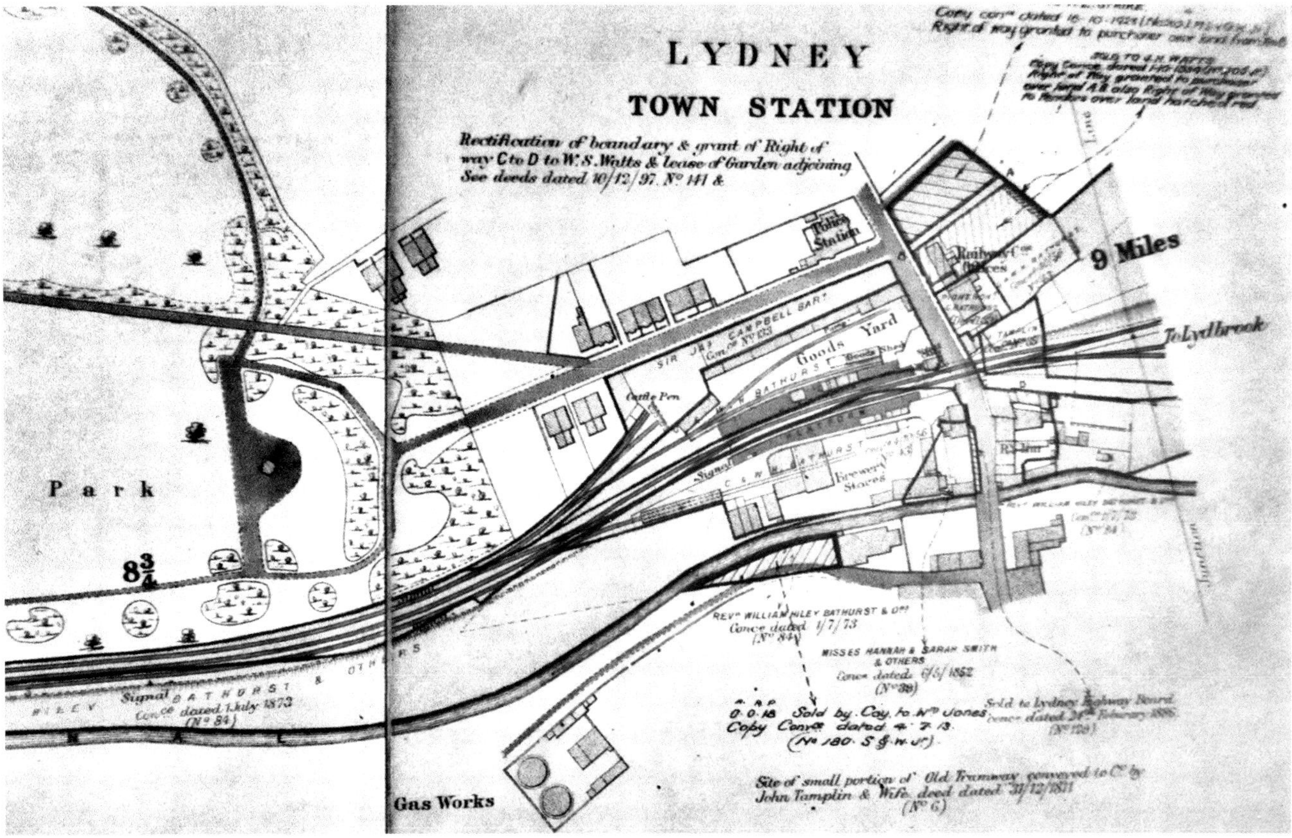

Lydney Town plan 1898.

The branch ran parallel with the GWR South Wales main line, became level at the 7½ mile post and made a junction at Otters Pool Junction signal box (7 miles 71 chains) which offered access to the South Wales main line. Beyond, it curved sharply to platforms at Lydney Junction (Severn & Wye) station (8 miles 15 chains) opened 16th October, 1879 and now the site of the southern terminus of the preserved Dean Forest Railway. This station opened for the Severn Bridge Railway, replacing the original terminus opened 23rd September, 1875 for the line from Drybrook Road.

The main building on the new Down platform incorporated recycled parts from the original terminal structure built by William Eassie & Company Limited. It was of red deal clad with 'rusticated weatherboarding' and lined with ¾-inch match boarding. The roof was covered with best Countess slates and ornamental ridge tiling. The small shelter on the opposite platform was a standard Gloucester Carriage & Wagon building.

The site of the original terminus was used for five carriage sidings and in 1880 a carriage shed was erected. Measuring 132 feet by 70 feet, it had formerly been a corrugated-iron church at Cheltenham and was purchased for £150, retaining its stained glass windows. Curiously it was not until 1897 that the Joint Committee's attention was brought to the fact that there were' no effective appliances for preventing vehicles running off the rails', so stop blocks were quickly erected at the end of each siding. The shed was dismantled in 1924 when the sidings were required for more sorting accommodation.

In 1910 three Severn Bridge Railway Sidings were authorised to be laid at a cost of £1,430 running from behind the Severn & Wye station to terminate just short of Lydney Junction signal box; they were known as Severn Bridge Nos. 3, 4 and 5.

A long footbridge connected the Severn & Wye station with that of the GWR main line. Originally having steps at each end which proved

Top: '43XX' class 2-6-0 No. 5398 passes Lydney Junction 14th June, 1958 with an Up goods. *R. E. Toop*

Centre: '14XX' class 0-4-2T No. 1401 heading a Berkeley Road – Lydney Town train being watered at Lydney Junction, 14th June, 1958. *R. E. Toop*

Right: A replica Severn & Wye Railway Lydney Junction station under construction by the Dean Forest Railway, 11th September, 1996. View towards the main line. *Author*

Lydney Junction, view towards the engine shed; the Severn & Wye platform is on the right.

M. E. J. Deane

Lydney Junction view Down. The Severn & Wye carriage shed, formerly a Cheltenham church, is the large building on the right. Milk traffic is quite heavy and the churns tidily placed. The flower bed is neatly edged with white-washed stones. The signal box in the distance on the left, was replaced in 1918.

Author's collection

Lydney Town: the footbridge at the rear of the signal box was built in 1904 at a cost of £567. The box operated the gates by means of a wheel. Beyond the crossing the horse and trap turning into Hill Street has come from Arnold Perrett's brewery stores. The man with a barrow is probably about to collect horse manure to put on his rhubarb.

Author's collection

inconvenient for prams and luggage barrows, in 1908 it was replaced at a cost of £2,700 by a ramped bridge approximately 240 yards in length and crossing 14 tracks.

One day about 1881, an Up Severn & Wye train stopped at Lydney Junction and a fair occupant in the story-teller's compartment – one of a pair of grown-up, respectable-looking young ladies journeying together – happened to look through the carriage window and espied the wooden bridge used for passenger communication with the South Wales Railway.

'Oh!' she exclaimed, 'is that the Severn Bridge?' 'I dare say it is,' was her companion's reply. 'Well,' said the first speaker, 'it isn't so large as I should have thought, but it is *very* pretty isn't it?' 'Yes it is,' said the other and after a few more enlightened criticisms, both these young ladies relapsed into their shilling novels and were oblivious when they actually passed over Keeling's great work. The absence of the Severn had not occurred to them!

From 21st May, 1955 the two Lydney stations were merged. The ex-Severn & Wye station closed officially on 30th November, 1964, but services had really ceased following damage to the Severn Bridge on 26th October, 1960. The Dean Forest Railway re-opened the former Severn & Wye station 2nd June, 1995 while the main line platforms have always remained in use.

Opposite the engine shed, on the west side of Church Road, was Richard Thomas & Son's Lydney Tinplate Works, established in 1816. In August 1941 the premises were taken over by the Admiralty as a munitions' store but returned to their owner in April 1946. The works closed in November 1957. Accessed from two sidings north of the Severn & Wye locomotive shed, rail inwards traffic consisted of chemicals, coal, steel bars and tin ingots, while tin-plated steel was despatched in vans. Initially shunted by horses, the Admiralty introduced locomotive working and this continued in the post-war period. Having no shed, the engine was kept in the yard. It was authorised to venture onto GWR and LMS tracks, but the company's steam crane, sometimes used for shunting, was not, yet occasionally trespassed. At various times wagon works existed in the triangular plot between the carriage shed and the two passenger stations.

Passing the engine shed, the double-track line was paralleled by a tramway until 1887 when it was replaced by a third track to Lydney Town. Near Lydney church was Pill House Bridge. When excavating for its foundation, 14 feet below the surface navvies found 32 cannonballs weighing 8-60 pounds each, together with part of an old boat, a teacup, a perfect bottle, pins and a tobacco pipe. As the soil comprised river mud, the footbridge bridge was built on a concrete foundation.

Lydney Town station (8 miles 73 chains), terminus of trains from Berkeley Road since 1929, was sited most conveniently in the centre of the town. It had two platforms and the goods yard had five sidings. On the Up platform was the original 1875 Eassie & Company timber-framed building. Adjacent on the same side was a siding serving a carriage shed between 1893 and 1907, while beyond a gate led to Lydney Foundry which specialised in making railway track components, but from 1892 was used as a brewery distribution depot by Arnold Perrett & Company; the siding was lifted on 26th July, 1933.

On the Down passenger platform stood a standard 1897 GWR red brick building. Adjacent was a cattle pen and end-loading dock, a Gloucester Wagon Company timber-framed goods shed, an engineer's stores and workshop, and a two-storey builders' merchant's warehouse.

At the Down end of the station complex was a GWR signal box which additionally worked the wheel-operated crossing gates. In response to the agitation of the local council regarding the delay caused by closed gates, a footbridge was provided in 1904.

The Lydney Harbour Branch

In the 1650s large ships were built at Lydney, but a few years later the River Severn had changed course placing the town inland. A 3 feet 8 inch gauge horse-drawn tramway to the river bank opened in October 1810 and when the outer harbour was finished in October 1821 the tramway was extended to serve it. The branch on the south side to the Lower Dock extended 1 miles 30 chains while that to the Upper Dock was 36 chains. When the broad gauge South Wales Railway was built in 1851, a level crossing was made over the tramroad and completed in just one day so as not to interfere unduly with tramroad traffic. At the crossing the gates were normally across the South Wales Railway, but opened at least five minutes before a South Wales train was due.

The broad gauge line was extended to the harbour in 1870 and a third rail was added in 1872 to carry traffic from the standard gauge mineral loop. Until broad gauge conversion 11th-12th May, 1872, the Severn & Wye operated broad, standard and tramway locomotives and tram horses.

The harbour had problems, suffering from a shortage of water in times of drought and a continual problem of silt, originally removed laboriously by hand. A steam dredger was provided in 1871.

In 1867 the harbour was handling pig iron, timber, bark and paving stone, but the principal proportion of the annual 200,000 tons was coal. A development at the harbour was Pine End plywood works.

On 1st January, 1950 control of the harbour was transferred from the British Transport Commission's Railway Executive to the Docks Executive, but coal traffic dwindled. The railway closed 18th November, 1960, the track on the south side of the harbour being lifted in the summer of 1962.

Lydney Harbour: tipping a coal wagon.
Author's collection

Chapter Eight

Locomotives

In March 1872 the Severn Bridge Company asked the Severn & Wye if it would work its line when completed. Severn & Wye Railway No. 5 *Forester* was the first locomotive recorded to have crossed the Severn Bridge, on 3rd September, 1879. It was an unusual machine having worked on no less than three gauges: firstly on a 3-feet 8-inch gauge tramway with flangeless wheels; then on the 7-feet ¼-inch broad gauge and then finally on the standard gauge.

Will Scarlet coupled to *Friar Tuck* tested the bridge on 15th September, 1879 and *Will Scarlet* drew 20 loaded wagons across on 2nd October, 1879. *Maid Marian* worked the first service train from Lydney-Sharpness. The Severn & Wye and Severn Bridge Railway livery was grass green with yellow lining.

Between October 1891 and 1895, MR 0-4-4Ts worked passenger trains from Gloucester–Lydney via the Severn Bridge.

In 1894 the MR imposed weight restrictions: the heaviest GWR locomotive allowed was a 'Dean Goods' 0-6-0 or a '2021' class 0-6-0T. A 'Dean' minus tender weighed 37 tons and a 2021 38 tons in working order. Light 'Johnson' 0-6-0s were used on MR ballast trains. After the 1923 Grouping, small ex-London & North Western locomotives were authorised, but probably never used.

In 1895 the GWR proposed that the locomotives and the locomotive department staff be abolished and engines hired from the parent companies, but the MR opposed this view and believed they should only be replaced when necessary.

In July 1895 the GWR proposed that it would provide locomotives for three years for:

Passenger trains 6½*d*. per train mile
Goods trains 8*d*. per train mile
Shunting 4*s*. per hour

and that any engine supplied by the MR would be subject to the same rates. This was agreed for a term of 10 years commencing 1st October, 1895.

Severn & Wye 0-6-0T *Friar Tuck* as MR No. 1122A. Built by Avonside in 1872, it was withdrawn in November 1911.
Author's collection

The existing stock of locomotives, carriages, goods stock and also machinery from Lydney Works, was valued and divided equally between the two companies. MR coaches would be almost solely used on the Berkeley branch and those of the GWR not used unless requested by the MR.

Most of the Severn & Wye engines acquired by the two companies were dispersed over their respective systems, but some of those on the GWR remained in the area for a few years. The GWR initially retained all the names except for *Wye*. Its name had been simply painted on its side and was covered over when the locomotive was repainted some time between 28th November, 1902 and 10th February, 1909. Conversely the MR removed all the names from its allocation of engines and by the time they had been in service with that company for a few years, they had been given a few Midland characteristics and at first glance appeared to be a standard 'Johnson' 0-6-0T, especially when they had acquired a dished Deeley smokebox door and chimney.

When the agreement that the GWR would supply motive power ended on 30th September, 1905, mileage rates were dropped in favour of an hourly charge of 5*s*. 6*d*.

In 1922 the GWR used '2361' class 0-6-0 Nos. 2365, 2374 and 2375 on coal trains to Taunton, these weighing 2½ tons more than a 'Dean Goods' and having larger cylinders and double frames. When, in September 1922, it was observed that these were contravening regulations, they were transferred elsewhere and replaced by 'Dean Goods' Nos. 2324, 2412 and 2574. On an unrecorded date between the Wars an 'Aberdare' 2-6-0 weighing 56 tons was inadvertently allowed to cross when working a diverted train.

In 1929 the possibility of using a Sentinel steam railcar between Lydney-Berkeley Road was considered, but the proposal turned down as it would not have been as economic, the roster requiring a locomotive which could be used for both passenger and goods working. In October 1932 consideration was given to working this service with an 'oil unit', but the idea was rejected due to difficulties which would be experienced with a tail load up the gradient of 1 in 132. It was suggested that a better proposition would be to

A '2021' class 0-6-0ST at Sharpness with a Down train of six-wheeled MR coaches. The locomotive has an extended smokebox, but an original length saddle tank. Cattle pens are on the right.
Author's collection

Class '4F' 0-6-0 No. 44264 comes off the Sharpness branch at Berkeley Road 11th May, 1961.

Author

introduce auto trains worked without a guard except for the 8.22 am Berkeley Road – Lydney and the 4.15 pm Lydney- Berkeley Road school trains which with about 60 children needed a guard. It was not until 30th November, 1936 that the auto-service was introduced.

The GWR Working Timetable Appendix for 1930 states that trains passing over the Severn Bridge must not be worked by more than one engine in front and that LMS passenger engines must not be allowed to pass over the bridge. The only engines permitted were:

GWR
2-4-0 Nos. 810, 3202/3, 3503, 3506/7, 3516.
0-6-0 Nos. 116, 132, 146, 331/3/4, 350/8, 363, 395, 601/7, 804, 934/7, 940/5, 1015, 1089, 1111, 1203, 1215, 2301-2360, 2381-2490, 2511-2580.
0-6-0T Nos. 2021-2160.

From 30th November, 1936 Lydney shed provided auto-fitted Collett 0-4-2Ts to work the new push-pull service.

In 1939 the GWR wanted to use the Severn Bridge as an alternative to the Severn Tunnel and proposed testing a '42XX' class 2-8-0T at 82 tons and '72XX' 2-8-2T at 92½ tons, but the LMS being in charge of the engineering side, banned the test. The 43½ ton Collett 0-6-0s were also refused, but under BR, with the bridge in Western Region control, from October 1950 the 62 ton '43XX' class 2-6-0s and Collett 0-6-0s were permitted to use the bridge.

The Working Timetable Appendix for December 1948 listed the only engines allowed over the bridge:

WR class '2301' (0-6-0) tender engines Nos. 2322-2356, 2382-2384, 2513-2579.
'2021' class (0-6-0T), '74XX' (0-6-0PT) and '14XX' (0-4-2T)
LMR '1P' 2-4-2T, '2P' 2-6-2T, '2F' LNWR 0-6-2T , '2F' Standard 2-6-0, '2F' LNWR 0-6-0 'Small Coal', '2F' LNWR 0-6-0 '18-inch', '2F' MR 0-6-0 Nos. 2987-3127, 3695, 22900-22984.

'54XX' class 0-6-0PT No. 5420 propelling the 08.15 to Lydney Town, 27th August, 1963.

Author

As from October 1950 '43XX' class 2-6-0s weighing 62 tons were allowed across the bridge working Sunday diversion trains and later Collett 0-6-0s were authorised.

The Working Timetable Appendix for June 1958 ruled that only these engines could work over the bridge:

Western Region: '43XX' 2-6-0 Yellow and Uncoloured groups.
London Midland Region: Class '2' 2-6-2T Standard; Class '2' 2-6-0 Standard; Class '2F' 0-6-0 Nos. 58115-58228.

BR Standard 'Class 4' 75XXX 4-6-0s and 'Class 4' 76XXX 2-6-0s to work with light tenders, 6 tons of coal and 3,500 gallons of water between Lydney Junction and Berkeley Road on running lines only. If an engine failed on the bridge, four wagons were required to be placed between it and the assisting engine. '28XX' and '38XX' 2-8-0s were permitted to work between Berkeley Road and Sharpness, but not across the bridge.

It was ruled: 'No engine or train must cross the iron part of the bridge in less than three minutes.' A restriction of a maximum speed of 40 mph was in force between Berkeley Road and Sharpness; 15 mph over the bridge and 25 mph Severn Bridge- Otters Pool Junction, Lydney.

In 1956 '64XX' and '16XX' 0-6-0PTs worked goods and passenger trains, though usually auto fitted 0-4-2Ts appeared on passenger duty. Elsewhere '16XX' were normally used as dock shunters rather than working passenger duties.

Although in 1950 the weight restriction meant that diverted Bristol-Cardiff expresses had to be hauled by 'Dean Goods' 0-6-0s, certainly about 1946 one of these veterans managed to reach about 55 mph on the main line. For the 5-coach 9.15 am Sunday train to Cardiff via the Severn Bridge, Bristol St Philip's Marsh used its best 'Dean Goods', cleaning its tubes and working it to Bristol Bath Road shed, (which dealt with passenger engines), ready for the Sunday working.

'72XX' 2-8-2T No. 7200 passes Lydney Junction 14th June, 1958 with a Down freight.

R. E. Toop

Although the Working Timetable Appendix stated that trains passing over the bridge must not be worked by more than one engine, it added that when an engine, or engine and brake van, was attached to a freight train working from Bristol or Stoke Gifford to Lydney or beyond via the Severn Bridge, in order to avoid occupation of the main line by light running, the engine, or engine and brake van, must be detached at Sharpness and follow the freight train independently from Sharpness to Lydney. Later this restriction was eased and it could be coupled in the rear.

Although in the British Railways' era there was but little variety of classes which worked over the bridge, quite a variety of numbers appeared. It was not the case of seeing the same engine again and again.

In the 1940s the 6.20 pm Lydney-Stoke Gifford and its 11.40 pm return service was one of the top link workings from Lydney shed, No. 2349 and No. 2350 heading the train regularly. The custom was to work tender-first over the bridge with a storm sheet helping to ameliorate the exposure and then use the turntable at Sharpness before going on to Stoke Gifford. While passing over the bridge in a gale it was not unknown for the weather sheet to be ripped from its moorings and coal blown off the fireman's shovel. From about 1950 the three 'Dean Goods' used for double-home turns from Lydney to Stoke Gifford, Weston-super-Mare and Taunton were replaced with '43XX' class 2-6-0s. When Moguls were allowed across the bridge, congestion in the Severn Tunnel could be eased by diverting some South Wales Cardiff-Bristol East and West depots via the bridge on a regular basis.

The maximum load up the 1 in 132 from Lydney to the Severn Bridge was 25 Class 1 loaded coal wagons. In the late fifties Bristol's St Philip's Marsh shed had one goods duty over the Severn Bridge. This was the 9.00 pm Stoke Gifford-Lydney. The outwards train could be up to 50 empty wagons. The engine was turned at Sharpness and proceeded tender-first to Lydney. There it took on water and hauled 24 loaded coal wagons to Sharpness where the load could be made up to 30 wagons. When this load was exceeded, and it frequently was, a banker was provided in the form of a 'Dean Goods' or

'2021' class 0-6-0PT. Reaching the bridge, the banker would then either return to Lydney, or follow at the rear to Sharpness and then double-head to Stoke Gifford. This happened when additional traffic, such as imported timber, was added to the train at Sharpness. This working was also used for returning empty acid tanks from the Lydney Tinworks to Avonmouth. When 2-6-0s took over this working, they had the disadvantage that they were too long for the Sharpness turntable and so were turned on the triangle at Berkeley Road before proceeding to Stoke Gifford.

A dedicated brake van was used for this working and branded: 'To work 6.20 pm Lydney to Stoke Gifford; 11.40 pm (SX) Stoke Gifford to Lydney; 11.50 pm (SO) Stoke Gifford to Lydney'.

An unusual working, probably in 1963 or 1964, was the appearance of BR Standard 'Class 2' 2-6-0 No. 78006 on a passenger train; as it was not fitted for auto-working, so necessarily had to run round its train.

With the end of steam, North British 'Type 2' diesel-hydraulic locomotives took over goods working Berkeley Road-Sharpness. BR/Sulzer Type 2 D5244 was tried on the Sharpness and Dursley branches in February 1968 before North British 'Type 2' diesels D6316 and D6320 were permanently based at Gloucester (Horton Road) depot for working the Sharpness, Forest of Dean, and Dursley branches, only returning to their parent depot, Bristol (Bath Road) for major servicing. These were replaced by 'Class 25' and later 'Class 31' and 'Class 37', with the occasional exciting appearance of a 'Peak Class 46'.

On 12th February, 1999 No. 37607 and 37610 were tried on the Berkeley – Hellafield nuclear flask train and proved successful; 25th February, 1999 was the last time English, Welsh & Scottish Railways were responsible for the run before Direct Rail Services took over the working. 'Class 20s' also appeared, but these were replaced by 'Class 68s', the first being Nos. 68001/026 on 8th March, 2018, while the first 'Class 88s' to appear were Nos. 88005 and 68003 on 8th October, 2019. As there is no run-round loop at Berkeley, trains have to run to Sharpness for this operation. The Royal Train hauled by No. 47500 was stabled on the branch overnight 2nd/3rd April 1980 and similarly No. 47792 and 47798 on 31st March and 1st April 1999.

68003 at the nuclear waste siding, Berkeley, spring 2022.
Author's collection

No. 37375 in the compound at Berkeley carrying the headboard: '7M56 1322 Berkeley to Sellafield last E. W. S. operated flask train 25th February, 1999'. *Richard Giles*

Sharpness Docks Steam Locomotives

With the construction of the New Dock at Sharpness 1870-4 the contractor used steam locomotives to haul the excavated spoil for removal by barge. The contractor connected his lines with the MR branch from Berkeley Road.

The Sharpness Docks and its railway were privately-owned until nationalised in 1948 when they became part of the Inland Waterways Executive of the British Transport Commission. It is not known what livery the locomotives wore when in private hands, but in BTC days they were painted black with red buffer beams and coupling rods and bore the Docks &

Sharpness Docks No. 3 0-4-0ST, an Avonside of 1902 awaiting disposal, 20th May, 1964. *Revd. Alan Newman*

Inland Waterways Executive emblem of a lifebelt and water on each side of the saddle tank or cab. The locomotive's number was either carried on a plate on the tank or cab side. Some were in white on the front buffer beam.

The dock company's engine placed traffic from the docks on the Joint Committee's sidings just outside the dock company's boundary fence. The Joint Committee's engine placed traffic to the dock company's line just inside the boundary fence.

Sharpness Docks Diesel Locomotives

DL 1 an 0-4-0 diesel-mechanical built by Ruston& Hornsby Limited left the factory for Sharpness on 30th March, 1961 and on arrival was placed in use. It had a 4 cylinder, 4-stroke direct injection diesel engine providing 88 bhp. Weighing 17 tons it could haul up to 510 tons; able to negotiate a 60 foot radius curve, it was particularly useful on North Dock lines. It stopped working about 1989 when rail traffic had declined owing to the type of goods imported and exported had declined and remaining traffic used the road. DL 1 was sold in 1997 and moved to Stoke Edith, Herefordshire.

Construction of DL 2 a Bagnall 0-6-0 diesel-mechanical No. 3151 was started in 1957 during Bagnall ownership and fitted with a Gardner 8L3 204 hp engine, but this was removed in March 1958. W. H. Dorman & Co. took over Messrs. Bagnall in 1959 and in July1961 the decision was made to install a Dorman 5QAT 208 hp engine fitted with flame-proofing as it was to become a demonstrator for British Petroleum, the locomotive being delivered to the BP Isle of Grain Kent Refinery in February 1962. Dorman & Co were taken over by English Electric in 1962 so on completion of the 8-month trial No. 3151 was sent to Robert Stephenson & Hawthorn's works at Darlington, another outpost of the English Electric empire. Here she was overhauled, the flame proofing removed together with the flanges on the centre driving wheels. In March 1963 No. 3151 was sold to the British Waterways Board for Sharpness Docks, arriving there in May.

DL 2 was unable to negotiate the tight curves on the North Dock lines and so was restricted to the south side of the docks. In November 1998 DL 2 was sold to the Appleby Frodingham Railway Preservation Society for running trains around some of the 90 miles of Scunthorpe Steel Works lines. In 2015 it was purchased by Kye Robinson and now works on the Lincolnshire Wolds Railway and is named *Debbie*.

British Waterways Board DL2 0-6-0DM shunts Cooper's Metals sidings with a load of rail and chairs at the weighbridge, Sharpness Dock, 18th April, 1967.
Richard Giles

Lydney shed: '8750' class 0-6-0PT No. 9727, left, shed plate 85B Gloucester Horton Road and '16XX' class 0-6-0PT No. 1630, also of 85B. The '16XX' class was a modernised version of the 1897 '2021' class, and first appeared nearly two years after the inauguration of British Railways.

Revd. Alan Newman

Lydney Locomotive Shed

Set north-west of the main line, it opened in 1868 as the main depot of the Severn & Wye Railway and adjoined the company's main works and repair shops. The three-road, stone-walled shed had timber ends to the roof gables and a pitched, slated roof. Each road was 11 feet in length and had an 84 feet-long pit. The adjacent repair shop opened in 1891, had only a single road, access being at the opposite end to that of the locomotive depot. It contained forges, lathes, including a wheel lathe, planing, shaping and drilling machines. The 82 feet 6 inch long inspection pit was straddled by a shear-legs but this was not used after 1895 and condemned as obsolete in March 1913. In pre-Joint Committee days, at times when its facilities were overwhelmed, engines were sent to Avonside Locomotive Works, Bristol for overhaul. Both buildings at Lydney were 115 feet in length, the shed being 37 feet wide and the repair shop 25 feet.

Water supply came from the River Lyd, pumped into two cylindrical tanks situated above the locomotive shed roof before being distributed to four columns: two outside the shed; one at the top of the yard and one at the west end of the Severn & Wye station.

The shed, originally coded LYD, in 1935 became a sub-shed to Gloucester, so its stock was coded GLO, in the BR era becoming a sub-shed of 85B (Gloucester). It closed 2nd March, 1964.

Lydney's Locomotive Allocation 1921

'2361' class 0-6-0 2365, 2374, 2375
'2021' class 0-6-0T 2024, 2025, 2029, 2032, 2040, 2041, 2043, 2053, 2068, 2069, 2080, 2084, 2085, 2087, 2093, 2138, 2150.
Total 20

'8750' class 0-6-0PT No. 3745, lacking a shed plate, at Lydney, 16th July, 1963, *Revd. Alan Newman*

Lydney's Locomotive Allocation 1934

'Dean Goods' 2349, 2428
'48XX' class 0-4-2T 4802
'2021' class 0-6-0T 2024, 2025, 2039, 2041, 2043, 2070, 2084, 2088, 2093, 2118, 2146, 2151, 2155, 2156, 2157
Total 18

Lydney's Locomotive Allocation 31st December, 1947

'Dean Goods' 0-6-0 2350
'14XX' class 0-4-2T 1409, 1456
'2021' class 0-6-0PT 2025, 2034, 2039, 2043, 2044, 2045, 2080, 2091, 2114, 2121, 2131, 2132, 2144, 2153, 2155, 2160.
Total 19

Lydney's Locomotive Allocation 27th February, 1954

'43XX' 2-6-0 (1)
'57XX' 0-6-0PT (2)
'16XX' 0-6-0PT (9)
'54XX' 0-6-0PT (2)
'14XX' (1)

Chapter Nine

Coaches

When the Severn Bridge Railway opened in 1879, the eight Severn & Wye coaches were insufficient for working traffic on both lines and by January 1880 the Bristol Wagon Company supplied three composites at £226 each and one third-class coach. In 1884 passengers grumbled at having to sit on hard, uncushioned seats which offered a much poorer ride when compared with the GWR and MR cushioned stock, so the Severn & Wye vehicles were suitably upholstered.

In 1890 four ex-Bristol Port Railway & Pier Company composites, (the BPRP had been taken over by the GWR and MR on 1st September, 1890) a third-class saloon and a passenger brake van were added to Severn & Wye stock.

The brake gear of the Severn & Wye stock was primitive, consisting of a chain wound on a drum adjacent to an axle. A friction clutch drove the drum from this axle so that when a lever in the brake van was pulled, a windlass drew the chain taut and thus applied the brakes. The Board of Trade, unhappy with this poor brake, insisted on an automatic system being fitted so the vacuum system was adopted at a cost of £825 16*s.* 2*d.* Fitting was carried out at Lydney under the supervision of a GWR employee from Swindon and brought into use on 29th May, 1891. At this date only four locomotives had been fitted, but the remainder were subsequently adapted and the system brought into general use on 20th May, 1892. Following the Bridge Company's take-over by the GWR and MR, the GWR supplied ten four-wheeled coaches, the MR supplying eight bogie vehicles. The latter were almost solely used on the Berkeley branch and the GWR stock only utilised at the GWR's request.

Certainly in 1932 the train service was operated using four LMS bogie and six GWR four-wheeled coaches; a half of each type being used while the remainder were being cleaned. In January 1934 three GWR low-roofed, brake-thirds Nos. 1965-7 were introduced. Originally built in 1897, during the First World War they were modified for use in ambulance trains, this involving the removal of guard's duckets, the fitting of gangway connections and Westinghouse brakes. In the post-war period they were returned to their original condition except the guard's look-outs were not replaced.

Severn & Wye saloon coach No. 12 after its repair in 1889.
Author's collection

A Down auto train propelled by '14XX' class 0-4-2T No. 1453 at Berkeley, September 1964.

D. Payne

51 feet ¾ inches long and 8 feet 6¾ inches wide, they were mounted on Dean's 10-feet suspension bogies. Similar in appearance to the 'Clifton Down' driving trailers built 1st August, 1897, they had five third-class compartments and a large luggage compartment with a pair of double doors each side.

In January 1934 their door locks were altered from the GWR standard to the LMS standard. This modification permitted the luggage compartment to be locked securely when the guard was not travelling, as some trains on the line ran without a guard. It is believed they were not equipped for auto-working and all three coaches were condemned in 1938.

Auto-working began on 30th November, 1936. It produced a saving in rolling stock as just three 60 feet trailers were required: one for regular trains; one for strengthening the school train while one was kept for cleaning or as a spare. Additionally auto-working saved daily 2½ hours of engine power and two guards.

Due to the fact that their locks were altered to the LMS standard, it is known that the following coaches worked over the Severn & Wye:

Clifton Downs driving trailers, Diagram A3:
No. 3332 September 1936-August 1939
No. 3337 August/September 1936-December 1938
No. 3339 September 1936-June 1939
No. 3340 September 1936-December 1940

In 1937 several Collett non-corridor brake-thirds were adapted for push-pull working. These were steel-panelled compartment driving trailers Diagram A33 (ex-Diagram D117 brake-thirds); half the £340 cost of conversion was paid by the LMS.

No. 4350 January 1937-post 1961
No. 4364 January 1937-post 1961

and steel-panelled 4-compartment driving trailers with a single large driver's window at the brake end Diagram A34 (purpose-built).

No. 1668 and No. 1671 worked over the line from April 1939 to post-1961. They were adapted to provide a 3 feet 3 inch driver's compartment at one end for push-pull working. At the luggage end of the coach the two small end windows were replaced by a large window in the end wall and a door fitted on each side. There was no door between the driver's and luggage compartments. They were fitted with ATC (Automatic Train Control) apparatus in 1946-7. These trailers were based at Lydney and branded 'To work between Lydney Town and Berkeley Road'. As was the usual practice with auto-working, to make conditions pleasanter for its crew, a locomotive usually hauled a trailer chimney-first with the dusty bunker behind; then on the return journey pushed the coach, which would have partly sheltered the bunker. Although at first glance A34 trailers were similar to those of Diagram A33, the former were purpose-built and had detail differences. The driver's compartment had a space of 4 feet 3½ inches between the partition and the end while the driver's door was flat, recessed and opened inwards.

Nos. 4350/64 were painted red-brown in 1942 and brown/cream in 1947 with a shield but no 'GWR' or 'Great Western'. At least one was repainted in carmine/cream and in 1956 one was seen in plain maroon.

GWR style saloon auto trailers were built by BR at Swindon in 1954 to Diagram A43 and at least one – W244W – appeared, though generally auto coaches on the branch were converted compartment type rather than saloons.

In 1955 BR Diagram A44 brake-third coaches were given an end window and converted for auto-working, at 4 feet 4 inches the driver's compartment was even larger.

Coaching stock exceeding 60 feet in length over the body and 9 feet 5 inches wide was prohibited from travelling across the bridge.

The chief civil engineer was to be consulted in all cases of loads producing axle weights in excess of 11½ tons; where axle spacing was 4 feet – 5 feet 6 inches if the axle weight exceeded 10½ tons and where the axle spacing was less than 4 feet.

'14XX' class 0-4-2T No. 1472 at the water crane, Sharpness, 25th August, 1964. *Revd. Alan Newman*

[No. 1.] 18

Circular Day-Trip Excursion Tickets will be issued Daily

(Sundays and certain other days, as shewn below, excepted)

From BRISTOL Stations, enabling passengers to visit

THE WYE VALLEY AND DEAN FOREST

Route No. 1.—Going via Severn Tunnel, Tintern, Monmouth, and Symond's Yat to Lydbrook Junction, and returning via Speech House Road, Lydney, Severn Bridge, Berkeley Road, and Charfield.
N.B.—Tickets will not be issued by this route on Saturday, Aug. 1st, Monday, Aug. 3rd, or Saturday, Aug. 8th.

Route No. 2.—Going via Charfield, Berkeley Road, Severn Bridge, Lydney, and Speech House Road to Lydbrook Junction, and returning via Symond's Yat, Monmouth, Tintern, and Severn Tunnel.
N.B.—Tickets will not be issued by this route on Saturday, Aug. 1st, Monday, Aug. 3rd, or Saturday, Aug. 8th.

Route No. 3.—Going via Severn Tunnel, Tintern, Monmouth, and Symond's Yat to Lydbrook Junction, and returning via Speech House Road, Lydney, Chepstow, and Severn Tunnel.
N.B.—Tickets will not be issued by this route on Saturday, Aug. 1st, or Monday, Aug. 3rd.

Route No. 4.—Going via Severn Tunnel, Chepstow, Lydney, and Speech House Road to Lydbrook Junction, and returning via Symond's Yat, Monmouth, Tintern, and Severn Tunnel.
N.B.—Tickets will not be issued by this route on Saturday, Aug. 1st.

Starting times for Routes I., III. and IV. — G.W.R. STATIONS.

FROM	AT a.m.	a.m.	a.m. (Aug. 3rd excepted.)	a.m. (August 3rd only.)
St. Anne's Park ..	f	...	9 26	9 26
Temple Meads ..	5 55	8 5	9 45	9 45
Lawrence Hill ..	6 0	8 10	9 22	9 50
Clifton Down ..	...	7 48	9 38	9 38
Redland ..	...	7 45	9 40	9 40
Montpelier ..	...	7 47	9 42	9 42
Stapleton Road ..	6 5	8 15	9 55	9 55
Ashey Hill ..	...	8 20	10 0	10 0

Starting times for Route II. — MID. STATIONS.

FROM	AT a.m.		a.m.
Temple Meads ..	7 5	...	10 5
St. Philip's ..	6 5	...	9 14
Clifton Down ..	...	...	9 16
Redland ..	...	...	9 18
Montpelier ..	...	..	9 20
Fishponds ..	7 14	...	9 28
Staple Hill ..	7 16	...	9 32

THIRD CLASS FARE for either circular trip.

3/3

Time-table of trains by which circular trip tickets will be available.

TRAINS TO AND FROM LYDBROOK JUNCTION via WYE VALLEY.

	a.m.	a.m.	a.m.	p.m.
Temple Meads (G.W.) dep.	5b55	8b5	9b45	...
Severn Tunnel Junc. arr.	6 33	8 53	10 27	...
" " ...dep.	6 55	10 5	10 40	..
Chepstow ... ... "	7 9	10 24	11 30	2 48
Tintern ... ... "	7 25		11 46	3 4
			p.m.	
Monmouth (Troy) ... arr.	7 46	...	12 7	3 25
" " ...dep.	9 30	...	12 30	3 38
Symond's Yat ... "	9 46	...	12 46	3 52
Lydbrook Junc. ... arr.	9 58	...	12 55	3 58

	a.m.	p.m.	p.m.	p.m. (July and Aug. only.)
Lydbrook Junc. ...dep.	10 52	3 12	5 14	...
Symond's Yat ... "	10 59	3 19	5 19	...
Monmouth (Troy) ... arr.		3 35	5 33	...
" " ...dep.	...	4 0	6 0	...
Tintern ... ... "	...	4 23	6 24	8 3
Chepstow ... ... "	...	4 42	6 40	8 21
Severn Tunnel Junc. arr.	...	4 57	6 55	8 33
" " ...dep.	...	5 10	7 17	9 13
Temple Meads (G.W.) arr.	...	6e5	8e0	10e0

TRAINS TO AND FROM LYDBROOK JUNCTION via FOREST OF DEAN.

	a.m.	a.m.	a.m.	p.m.
Temple Meads (G.W.) dep.	5b55f	8b5	9b45	...
Severn Tunnel Junc. arr.	6 33	8 53	10 27	...
" " ...dep.	7 17	10 5	11 5	...
Chepstow ... ... "	7 37	10 24	11 16	...
Lydney (G.W.) ... arr.	7 55	10 40	11 30	...
Temple Meads (Mid.) dep.	7b5	...	10b5	...
Berkeley Road ... arr.	8 3	...	10 45	...
" ...dep.	8 50	...	11 15	2 19
Berkeley ... ... "	8 57	...	11 21	2 25
Sharpness ... ... "	9 3	...	11 27	2 31
Severn Bridge ... "	9 9	...	11 32	2 37
Lydney Junc. ... ... arr.	9 16	...	11 39	2 44
Lydney Junc. ... ...dep.	9 21	...	11 46	3 0
Lydney Town ... ... "	9 24	...	11 49	3 3
			p.m.	
Speech House Road ... "	9 46	...	12 11	3 19
Cinderford ... ... "	9 57	...	12 22	3 30
Lydbrook Junc. ... arr.	10 20	...	12 45	3 52

	p.m.	p.m.	p.m.	p.m.	p.m.	p.m.	
Lydbrook Jc. ...dep.	12 28	...	4 2	...	..	6 25	...
Cinderford ... "	12 57	...	4 24	...	6 0	6 48	...
Speech House Rd. "	1 7	...	4 35	...	6 11	7 0	...
Lydney Town ... "	1 30	2 53	4 55	...	6 31	7 20	...
Lydney Junc. ... arr.	1 32	2 55	4 57	...	6 33	7 23	...
Lydney Junc. ...dep.	1 34	2 56	5 15	...	6 35	7 25 (July and August only.)	
Severn Bridge ... "	1 41	3 3	5 22	...	6 42	7 32	
Sharpness ... "	1 47	3 9	5 28	...	6 55	7 38	
Berkeley ... "	1 53	3 15	5 34	...	7 1	7 44	
Berkeley Road arr.	1 58	3 20	5 40	...	7 10	7 50	
" ...dep.	2 13	4 8	6 2	...	8 13	8 13	
T. Meads (Mid.) arr.	3 8	5d 3	7d 1	...	9d11	9d11	
Lydney (G.W.) dep.	2 23	4 10	...	...	6 48	8 10	...
Chepstow ... "	2 36	4 42	...	6 40	7 5	8 29	...
Severn Tunl. Jc. arr.		4 57	...	6 55	7 21	8 48	...
" dep.	...	5 10	...	7 17	9 13	9 13	...
T. Meads (G.W.) arr.	...	6e5	...	8e0	10e0	10e0	...

b For times of leaving other Bristol stations see top of page.

d See next page. *e* These trains also connect to Ashley Hill, Stapleton Road, Montpelier, Redland, Clifton Down, Lawrence Hill, and St. Anne's Park. *f* Not run on Monday, Aug. 3rd.

(Continued on page 19.)

For Special Notices relating to the issue of Excursion Tickets, see page 2.

Booklet of excursions from the Bristol area, summer 1908.
Note that a station name has been misspelt.

Wye Valley and Dean Forest Circular Trips *(continued).*

d These trains also connect to Staple Hill, Fishponds, Montpelier, Redland, Clifton Down, St. Philip's, and St. Anne's Park. Passengers returning *via* Berkeley Road and Charfield to Montpelier, Redland, and Clifton Down must change at Fishponds. The Tickets are not available *via* Temple Meads; such passengers may, however, travel to Temple Meads and terminate the journey there, giving up their Tickets at Fishponds.

BREAK OF JOURNEY.—Passengers may break the journey at either of the intermediate Stations shewn in the foregoing time-table, and they may leave the railway at any station and go by road to proceed by a subsequent train from any station in advance on the route for which the tickets are available, but the circular trip must be completed on the same day.

Tickets issued by	At	Will be available to return to
Either Co.	Clifton Down	Clifton Down or Temple Meads.
Do.	Redland	Redland or Temple Meads.
Mid. Co.	Montpelier	Montpelier (G. W.), or Ashley Hill (G. W.)
Do.	Temple Meads	Lawrence Hill (G. W.), Temple Meads (G. W.) or St. Anne's Park (G. W.)
Do.	St. Philip's	Do. Do.
Do.	Fishponds	Stapleton Road (G. W.)
Do.	Staple Hill	Do. [Temple Meads
G. W. Co.	St. Anne's Park	Temple Meads (Mid.) or St. Anne's Park, *via*
Do.	Temple Meads	Temple Meads (Mid.).
Do.	Lawrence Hill	Do.
Do.	Stapleton Road	Fishponds (Mid.), Montpelier (Mid.), or Temple Meads (Mid.)
Do.	Montpelier	Montpelier (Mid.), or Temple Meads (Mid.)

DAILY (Sundays and Saturday, August 1st, excepted), Day Excursion Tickets will be issued to

THE FOREST OF DEAN

(Via SEVERN TUNNEL and LYDNEY).

LEAVING	To Chepstow, Lydney, Severn Bridge, Sharpness, Coleford, Speech House Road, Cinderford, or Lydbrook Junction.			To Chepstow and Lydney.	To Speech House Rd. and Cinderford.
	Monday Aug. 3 only.	Daily, Mon., Aug. 3, excepted.		Daily.	Daily in J'ly & Aug. & on Sats. in Sept.
	A	A	A	B	B
	A. M.	A. M.	A. M.	P. M.	P. M.
St. Anne's Park	9 26	...	9 26	...	...
Temple Meads ...	9*45	5 55	9 45	1 15	1 15
Lawrence Hill ...	9*50	6 0	9 22	1 20	1 20
Clifton Down ...	9 38	...	9 38	1 12	1 12
Redland	9 40	...	9 40	1 14	1 14
Montpelier ...	9 42	...	9 42	1 16	1 16
Stapleton Road...	9*55	6 5	9 55	1 25	1 25
Ashley Hill ...	10* 0	...	10 0	1 29	1 29

Return Fares, Third Class.

To Chepstow or Lydney. By trains marked A	To Chepstow or Lydney. By trains marked B	To Severn Bridge or Sharpness.
2/3	2/-	2/3

To Speech House Road and Cinderford. By trains marked A	To Speech House Road and Cinderford. By trains marked B	To Coleford or Lydbrook Jct.
2/9	2/-	2/9

* Through train to Lydney, for Sharpness, Speech House Road, &c.

TIMES OF RETURN TRAINS.

	Daily in July, August and September.				
	P. M.	P. M.	P. M.	P. M.	P. M.
Lydbrook Junction .. ...dep.	...	4 2	...	...	6 25
Cinderford... ... ... ,,	...	4 24	...	6 0	6 48
Speech House Road... ... ,,	...	4 35	...	6 11	7 0
Coleford ... ... ,,	...	4 15	..	5 55	...
Sharpness ... ...dep.	4 23	...	6 27	...	...
Severn Bridge ... ... ,,	4 28	..	6 32	...	...
Lydney Junction ... ...arr.	4 35	4 57	6 42	6 33	7 23
Lydney, G.W.R. ... ...dep.	6 48				8 10
Chepstow ... ...arr.	7 5				8 28
Do. (depart for Bristol) ...	8 29				8 29

Passengers can break the journey at any of the stations mentioned above, intermediate between the points to and from which the tickets are available, if the train service permits.

Passengers change at Severn Tunnel Junction and Lydney (G. W. R.) in both directions unless otherwise shown.

NOTE.—Passengers visiting Symond's Yat can return either from Lydbrook Junction or Coleford (S. & W.) Stations.

For Special Notices relating to the issue of Excursion Tickets, see page 2.

BERKELEY ROAD, SHARPNESS, LYDNEY, COLEFORD, and LYDBROOK.—S. & W., S. B., & M.

Temple Mead Sta.,	mrn	mrn	mrn	mrn	mrn	aft	aft	aft	aft	aft
Bristol 266 dp	...	7 15	....	9 10	1015	1 15	3 20	....	5 0	7 0
265 Gloucester "	5 15	7b48	8 10	....	1045	1 18	3 15	4 46	4 46	7 15
Berkeley Road dep	5 38	8 18	8 52	1010	1117	2 12	4 15	5 20	5 59	8 0
Berkeley	5 45	8 23	8 57	....	1122	2 18	4 21	5 25	6 5	8 5
Sharpness ar	5 50	8 28	9 2	1019	1127	2 23	4 26	5 30	6 10	8 10
(New Station) dp	Mondays only. a	8 29	9 3	1020	1130	2 24	4 27	Stop	6 12	8 12
Severn Bridge.* ..		8 34	9 8	1025	1135	2 29	4 33		6 17	8 17
Lydney Jn. 34, 35a		8 41	9 15	1032	1142	2 36	4 40		6 24	8 24
Newport 34 ar		....	1050	....	1255	5 25	....		7 49	1022
Cardiff 34 "		....	1114	....	1 23	5 53	....		8 15	....
Lydney Junc. dep	mrn	8 53	9 22	1035	1144	2 37	4 41	4 45	6 27	8 25
Lydney Town	7 30	8 56	9 25	1038	1147	2 40	4 44	4 50	6 30	8 28
Whitecroft	7 37	9 3	9 32	....	1154			4 57	6 38	....
Parkend	7 40	9 8	9 36	1048	1158			5 2	6 42	....
Milkwall	8 25	....	9 55	Mons. only.	1216			5 16	....	....
Coleford † 36 arr	8 30	....	10 0		1218			5 20	....	....
36 Monmouth ‡ ar	....	....	1210		2 35			5 50	....	....
Coleford ....dep	....	8 45	....		9 20			....	6 0	....
Speech House Road	7 50	9 16	9 50	1058	12 8			....	6 50	....
Cinderford	8 3	9 25	....	11 7	1218			....	7 0	....
Drybrook Road	8 6	9 28	....	1110	1220			....	7 3	....
Upper Lydbrook	8 20	9 38	....	1120	1230			....	7 12	....
Lower Lydbrook	Sig.	Sig.	....	....	Sig.			....	Sig.	....
Lydbrook J. 32, 33	8 25	9 48	....	1148	1240			....	7 18	....
Monmouth 33 a	8 52	1221	....	1221	3 23			....	7 50	....

b Except Mondays. * Station for Blakeney. † Station for Staunton. ‡ Troy Station.

May Hill Station,	mrn	mrn	mrn	mrn	aft	aft	aft	aft	aft	aft
32 Monmouth..dep		....	....	9 34	....	1229	3 59	3 59	....	7 10
Lydbrook Junc. dep		....	8 55	1215	....	3 17	5c50	4 25	....	7 35
Lower Lydbrook		....	Sig.	Sig.	....	Sig.	....	....	....	Sig.
Upper Lydbrook		....	9 4	1225	....	3 27	5c59	4 34	....	7 44
Cinderford		....	9 18	1242	....	3 45	6c10	....	....	8 0
Drybrook Road		....	9 21	1246	....	3 47	....	....	....	8 4
Speech House Road		....	9 28	1251	....	3 54	6 24	4 45	7a0	8 13
Coleford † 36..arr		....	10 0	....	....	5 20	....	Mondys only a	....	....
36 Monmouth ‡.dp		8 0	8 0	9 20	....	1250	4 45		....	....
Coleford .....dep		8 45	9 20	1245	....	3 45	6 0		....	....
Milkwall		8 48	9 25	1250	....	3 49	6 8		....	....
Parkend		Stop	9 36	1 3	....	4 6	6 32		....	8 23
Whitecroft	mrn		9 39	1 7	....	4 9	6 36	aft	....	8 27
Lydney Twn [34	7 30	....	9 47	1 14	3 15	4 16	6 43	6 47	7 20	8 35
Lydney J. 35, a	7 32	mrn	9 50	1 17	3 18	4 18	6 45	6 50	7 23	c 8 37
35 Cardiff dp	....	1010	7 0	1010	1228	....		4 5	6 10	
35 Newport "	....	1030	7 25	1030	1248	....		4 30	6 30	
Lydney Jn. dp	7 33	1142	10 0	1 19	3 19	4 26		6 52	7 25	c Mondays only.
Severn Bridge*	7 41	....	10 7	1 26	3 26	4 33		6 59	7 31	
Sharpness ar	7 47	1152	1012	1 32	3 32	4 39	aft	7 5	7 38	
(New Sta.) dp	7 49	....	1019	1 35	3 34	4 45	5 37	7 7	7 39	
Berkeley. .[266	7 51	....	1024	1 41	3 39	4 51	5 42	7 11	7 44	
Brkely Rd 266,	8 0	....	1032	1 46	3 43	4 56	5 47	7 19	7 49	
266 Gloucestr a	8 45	....	1124	2 35	4 31	6 35	6 35	7 56	9 25	
265 Bristol.. "	9 42	....	1140	2 43	4 52	5 58	....	....	8 50	

a For Pleasure Parties when required; previous notice to be given. [277

August 1887.

BERKELFY ROAD, LYDNEY, and LYDBROOK (1st and 3rd class).—Great Western and Midland Joint.

Traffic Man., John A. Carter, Lydney.

Week Days.

Miles.	Temple Meads Sta.,	mrn	mrn	mrn	mrn	aft	aft	aft	aft	aft	aft
	578 Bristol .....dep.	....	7 5	8 5	10 5	1 5	....	2 35	5 5	7 20	Thursdays & Saturdays.
	573 Gloucester.. "	....	....	8 0	1045	1 35	....	3 28	5 25	7 35	
	Berkeley Road ..dep.	....	8 10	8 50	1115	2 19	....	4 10	6 12	8 20	
2¼	Berkeley	....	8 15	8 57	1121	2 25	3c35	4 17	6 19	8 27	
4	Sharpness	....	8 20	9 3	1127	2 31	3 45	4 23	6 27	8 33	
5¼	Severn Bridge*	....	8 26	9 9	1132	2 37	3 50	4 28	6 32	8 38	
8	Lydney Junc. 58, 62	....	8 32	9 16	1139	2 44	3 57	4 35	6 42	8 45	
33	58 Newport ¶ .arr.	....	....	1056	1249	....	5 15	....	8 5	1016	
44¾	58 Cardiff ‖ .... "	....	....	1116	1 23	....	5 45	....	8 3	1040	
—	58 Gloucester .dep.	....	7 45	7 45	11 0	1 53	....	3 25	6 17	7 25	9 0
—	Lydney Junc....dep.	....	8 34	9 21	1145	2 46	....	4 40	6 53	8 46	9 40
8¾	Lydney Town	7 20	8 36	9 24	1148	2 48	....	4 43	6 56	8 48	9 43
11	Whitecroft	7 27	...	9 31	1155	....	....	4 50	7 3	....	9 50
12¾	Parkend.........arr.	7 29	....	9 33	1157	....	....	4 52	7 5	....	9 52
14	Milkwall †	7 50	....	9 52	1215	....	....	5 11	7a20	....	....
15	Coleford ‡ 67 ..arr.	7 54	....	9 56	1219	....	....	5 15	7a24	....	....
21	67 Monmouth § arr.	8 45	....	1156	2 45	....	....	5 47	....	....	....
—	Parkend.........dep.	7 30	....	9 35	1159	....	....	4 55	7 6	....	9 53
14¾	Speech House Road	7 44	....	9 46	12 9	....	....	5 5	7 13	....	10 0
17¼	Drybrook Road	7 53	....	9 53	1216	....	....	5 12	d	....	10 7
18½	Cinderford .... arr.	7 58	....	9 57	1220	....	....	5 16	7 23	....	1011
	(New Station) dep.	8 7	....	10 1	1223	....	....	....	7 26	....	....
—	Drybrook Road	8 11	....	10 5	1227	....	....	....	7 30	....	....
18¾	Upper Lydbrook	8 20	....	1014	1236	....	....	....	7 39	....	....
20¼	Lydbrook Junc. 98 ar	8 26	....	1020	1245	....	....	....	7 45	....	....
27¼	98 Monmouth ** arr.	8 49	....	1121	3 30	....	....	....	....	....	....

Miles.	May Hill Station,	mrn	mrn	mrn	mrn	mrn	mrn	aft	aft	aft	aft	aft
	98 Monmouth ..dep.	....	....	....	7b34	....	9 34	....	3 42	....	7 42	....
	Lydbrook Junc..dep.	....	....	....	8 40	....	1228	....	4 2	....	8 15	....
1¼	Upper Lydbrook	....	....	....	8 46	....	1238	....	4 8	....	8 21	....
5¼	Drybrook Road	....	....	....	....	....	1247	....	....	....	8 31	....
6½	Cinderford.... arr.	....	....	....	8 58	....	1251	....	4 21	....	8 35	....
	(New Station) dep.	....	....	8 18	9 1	....	1254	....	4 24	6 0	8 38	1015
—	Drybrook Road	....	....	8 22	9 5	....	....	....	4 28	6 4	8 42	1019
6¾	Speech House Road	....	....	8 29	9 12	....	1 4	....	4 35	6 11	8 50	1026
8¼	Parkend.........arr.	....	....	8 38	9 21	....	1 12	....	4 44	6 20	8 59	1032
—	67 Monmouth § dep.	....	....	....	7 58	....	9 40	....	1235	4 45	....	....
—	Coleford ‡.....dep.	....	....	8 13	8 55	....	1247	....	4 15	5 55	8a34	....
—	Milkwall †	....	....	8 17	8 59	....	1251	....	4 19	5 59	8a38	....
—	Parkend.........dep.	....	....	8 39	9 28	....	1 14	....	4 45	6 21	9 0	1033
9¼	Whitecroft	....	....	8 41	9 31	....	1 17	....	4 47	6 24	9 3	1035
11¾	Lydney Town	7 25	8 5	8 48	9 38	....	1 25	2 53	4 55	6 31	9 10	1042
12¼	Lydney J. 58, 62 arr.	....	8 7	8 50	9 40	....	1 27	2 55	4 57	6 33	9 12	
31¼	62 Gloucester .arr.	....	....	9 34	1125	....	2 33	....	5 40	7 55	9 58	Thursdays & Saturdays.
—	62 Cardiff ‖ ...dep.	....	6 30	....	7 54	1023	....	1 25	4 0	4 25	....	
—	62 Newport ¶. "	....	6 50	....	8 20	1048	....	1 53	4 22	5 0	....	
—	Lydney Junc....dep.	....	8 8	....	9 42	1150	1 29	2 56	5 10	6 35	....	
15¼	Severn Bridge*	7 35	8 15	....	9 49	1157	1 36	3 3	5 17	6 42	....	
16½	Sharpness	7 42	8 21	....	9 56	12 2	1 42	3 9	5 23	6 51	....	
18¼	Berkeley.......[578	7 48	8 27	....	10 2	12h8	1 48	3 15	5 29	7 0	....	
20¼	Berkeley Road 573,	7 54	8 32	....	10 8	....	1 53	3 20	5 35	7 5	....	
35¼	578 Gloucester..arr.	8 42	9 0	....	1053	....	2 32	4 5	6 45	7 53	....	
42¼	573 Bristol ‡‡ .. "	9g2	9 42	....	1140	....	3 10	5 3	7 5	9 15	....	

NOTES.

- a Thursdays and Saturdays.
- b Wednesdays only.
- c Runs from Berkeley on notice being given to the Station Master not later than 2 aft.
- d Stops when required on informing the Guard or Station Master.
- g Mondays, Tuesdays, & Thursdays.
- h Runs forward to Berkeley on notice being given to the Guard.
- * Station for Blakeney.
- † Station for Clearwell.

‡ Station for Staunton. § Troy Station. ‖ Great Western Station. ¶ High Street Station. ** May Hill Station. ‡‡ Temple Meads.

April 1910.

BERKELEY ROAD, LYDNEY, and LYDBROOK.—Severn and Wye—Great Western and Midland.

Week Days only.

Miles.		mrn	mrn	mrn	aft	aft	aft	aft	aft	aft
	630 Bristol (Temple Meads) dep.	....	7 40	1035	....	1 18	2 30	4 57	6 37	Saturdays only.
	626 Gloucester (Midland).. "	....	7 50	1054	....	1 50	3 42	5 42	7 7	
—	Berkeley Road ....dep.	....	8 50	1145	....	2 40	4 18	6 35	7 53	
2¼	Berkeley	....	8 56	1151	....	2 46	4 24	6 41	7 58	
4	Sharpness	....	9 2	1157	....	2 52	4 30	6 49	8 3	
5¼	Severn Bridge, for Blakeney ...	....	9 7	12 2	....	3 2	4 35	6 57	8 8	
8	Lydney Junction 70, 71 ...arr.	....	9 15	12 9	....	3 9	4 42	7 6	8 15	
33	70 Newport ....arr.	....	1049	2 10	....	5A6	6 16	9 11	10 1	
44¼	70 Cardiff (G. W.) .... "	....	1110	2 32	....	5A33	6 40	9 45	1025	
—	70 Gloucester (G. W.).....dep.	....	8 44	9 16	....	....	3 20	6 10	7 20	
—	Lydney Junction ....dep.	....	9 23	1212	....	....	4 46	7 12	8 17	....
8¾	Lydney Town	7 20	9 26	1215	....	....	4 48	7 14	8 20	8 23
11	Whitecroft	7 27	9 35	1223	....	....	4 56	7 24	....	8 32
12¼	Parkend ....arr.	7 30	9 38	1226	....	....	4 59	7 27	....	8 35
14	Milkwall, for Clearwell	7 51	9 56	1243	....	....	5 18	7 51	....	....
15	Coleford, for Staunton ...arr.	7 55	10 0	1247	....	....	5 22	9 55	....	....
—	Parkend ....dep.	7 31	9 40	1228	....	....	5 2	7 34	....	....
14¾	Speech House Road	7 46	9 51	1237	....	....	5 12	7 47	....	....
17¼	Drybrook Road	7 55	9 59	1245	....	....	5 22	7 56	....	....
18½	Cinderford 74 .... arr.	7 59	10 3	1249	....	....	5 26	8 1	....	....
	dep.	8 4	10 8	....	1 48	....	5 31	8s6	....	....
—	Drybrook Road	8 8	1012	....	1 52	....	5 35	8s10	....	....
18¾	Upper Lydbrook	8 17	1021	....	2 1	....	5 44	8s20	....	....
20¼	Lydbrook Junction 72 .....arr.	8 23	1027	....	2 7	....	....	8s26	....	....
27¼	72 Monmouth (May Hill) ...arr.	8 52	1125	....	3 35	....	....	....	....	....

Miles.		mrn	mrn	mrn	mrn	non	aft	aft	aft	aft	aft	aft	aft
	72 Monmouth (May Hill)....dep.	....	....	....	....	....	....	12 34	....	3 46	....	....	8 29
—	Lydbrook Junction ....dep.	....	....	8 45	....	....	....	1 25	....	4 5	....	....	8 50
1¼	Upper Lydbrook	....	....	8 51	....	....	....	1 31	....	4 11	5 55	....	8 56
5¼	Drybrook Road	....	....	9 0	....	....	....	1 40	....	4 22	6 5	....	9 5
6½	Cinderford 74 .... arr.	....	....	9 3	....	....	....	1 44	....	4 25	6 9	....	9 9
	dep.	....	....	9 7	....	....	1 20	Stop	....	4 29	6 12	8e20	9 15
—	Drybrook Road	....	....	9 11	....	....	1 24		....	4 33	6 16	8e24	9 19
6¾	Speech House Road	....	....	9 18	....	....	1 31	....	....	4 40	6 23	8e31	9 26
8¼	Parkend ....arr.	....	....	9 27	....	....	1 39	....	....	4 49	6 34	8e41	9 37
—	Coleford ....dep.	....	8 10	9 0	....	12 0	....	....	4 7	....	6 0	8e15	9 10
—	Milkwall, for Clearwell	....	8 14	9 4	....	12 4	....	....	4 11	....	6 4	8e19	9 14
—	Parkend ....dep.	....	8 33	9 30	....	1222	1 41	....	4 26	4 51	6 35	8 42	9 39
9¼	Whitecroft	....	8 36	9 34	....	1225	1 44	aft	4 28	4 55	6 38	8 45	9 43
11¾	Lydney Town	7 5	8 43	9 41	1038	1235	1 52	2 50	4 36	5 5	6 46	8 53	9 51
12¼	Lydney Junction 70, 71 ....arr.	7 7	8 46	9 44	1040	1237	1 54	2 52	4 39	5 7	6 48	....	....
31½	71 Gloucester (G. W.)......arr.	8 40	9 35	1035	1135	1 55	2F41	3 43	5 42	6 45	7 44	....	Saturdays only.
—	71 Cardiff (G. W.) ....dep.	....	....	....	7 45	....	1150	12F55	....	4 3	4 35	....	
—	71 Newport .... "	....	....	....	8 13	....	1220	1F23	..	4 24	5 5	....	
—	Lydney Junction...........dep.	7 8	....	....	1042	....	1 56	2 54	....	5 12	6 51	....	
15¼	Severn Bridge, for Blakeney	7 15	....	....	1049	....	2 3	3 1	....	5 19	6 59	....	
16¼	Sharpness	7 22	....	....	1054	....	2 9	3 7	....	5 25	7 5	....	
18¼	Berkeley	7 28	....	....	11 0	....	2 15	3 13	....	5 33	7 12	....	
20½	Berkeley Road, 626, 630 ...arr.	7 34	....	....	11 6	....	2 21	3 18	....	5 39	7 18	....	
35½	630 Gloucester (Midland)...arr.	8 25	....	....	1216	....	....	4 22	....	6 37	8 0	....	
42½	626 Bristol (Temple Meads). "	9 50	....	....	12 0	....	3 30	4 52	....	7 30	8 50	....	

A Arrives Newport at 4 20 aft. and Cardiff 4 25 aft. on Fridays. e Except Saturdays. F Fridays only. s Saturdays only.

July 1922.

Chapter Ten

Timetables and Train Working

From November 1879 Bradshaw's timetables combined the Severn & Wye and the Severn Bridge timetable with that of the MR's Sharpness branch, although the latter remained MR property until it became part of the joint line.

Initially a service of seven trains daily was operated, the first and last being the fastest, taking 25 minutes Down and 24 minutes Up; additionally two short workings each way were run from Sharpness-Lydney Junction. In the early days, a five-coach train left Berkeley Road and on reaching Coleford Junction in the Forest of Dean, the rake was divided. The main train, consisting of the first three coaches, proceeded to Cinderford and Lydbrook, while the last two were left to be drawn in reverse up the gradient of 1 in 30 to Coleford. On the return journey, the two Coleford coaches were attached at Coleford Junction to the rear of the Cinderford and Lydbrook-Berkeley Road train. Thus there was a 3/2 and 2/3 division, the centre coach alternating between Coleford and Lydbrook. The timetables for trains between Berkeley Road and Lydney Junction varied little over the years, with seven to eight trains each way, weekdays-only and taking approximately 23 minutes for the journey Berkeley Road-Lydney Junction.

By August 1887 the service offered eight Down trains plus an early morning Mondays-only from Berkeley Road-Sharpness and an every day early evening run Berkeley Road-Sharpness. Seven Up trains ran, plus a morning one from Lydney Junction-Sharpness and an afternoon train from Sharpness-Berkeley Road. Although generally throughout the life of the line passenger trains called at all stations, in 1887 the 10.10 am Down omitted Berkeley, and the Lydney Junction-Sharpness omitted Severn Bridge.

From August 1888 the MR ran a four-coach train from Gloucester to Parkend in the Forest of Dean via the Severn Bridge; en route two Severn & Wye coaches for Coleford attached to its rear were uncoupled at Lydney Junction. It is believed that tank engines were used initially, but by 1889 a 4-2-2 appeared on this working.

In the 1880s the MR on Mondays and Fridays issued excursion tickets from Cheltenham, Gloucester, Stonehouse, Dursley and Cam, to Severn Bridge, Lydney, Speech House Road, Coleford and Lydbrook Junction. Cheap return tickets were offered any day to pleasure parties of no less than six first, or ten third-class passengers.

By October 1891 the MR passenger train from Gloucester to Lydney via the Severn Bridge worked by a MR 0-4-4T, left Berkeley Road at 4.02 pm, ran to Coleford and then returned at 6.10 pm, but the service was withdrawn in 1895. E. L. Ahrons in *Locomotive & Train Working in the Latter Part of the Nineteenth Century* writes that MR 0-4-4Ts Nos. 1280/1 were used on this service.

Certainly in 1893 the MR ran trips from Birmingham to Ilfracombe using rail from Birmingham-Sharpness and then a paddle steamer to Ilfracombe.

The service for the first four months of 1902 offered six Down (one mixed) plus one Sharpness-Lydney Junction and five Up (one mixed) plus one from Severn Bridge-Berkeley Road and another Lydney Junction-Sharpness.

In April 1910 there were seven Down trains, plus the 3.45 pm Sharpness-Lydney Junction which could start at 3.35 pm from Berkeley if notice was given to the station master no later than 2.00 pm that day. Six Up trains were operated, plus one from Severn Bridge-Berkeley Road and another Lydney Junction-Sharpness, the latter extending to Berkeley on notice being given to the guard.

From 7th January, 1923 Bristol–South Wales trains were diverted over the bridge on winter Sundays to allow the engineering department possession of the Severn Tunnel.

Until 6th July, 1929 two trains daily ran Lydbrook Junction-Berkeley Road; two from Cinderford-Berkeley Road and five Lydney Town-Berkeley Road. In the Down direction two ran Berkeley Road-Lydbrook Junction; two Berkeley Road-Cinderford, three Berkeley Road-Lydney Town and one from Berkeley Road-Lydney Junction and Cardiff. As settlements in the Forest of Dean were small, traffic was light and this was the reason that most passenger services over the Severn & Wye were withdrawn. Remaining trains were fairly well patronised until the bridge was struck in 1960. From 1st January, 1934 first-class was withdrawn and trains run without guards.

On 30th November, 1936 push-pull operation had been instituted and trains were one-class only, seven down, plus one Saturdays-only and one Monday-Friday during the school term: the Up service was the same. By October 1941 wartime had reduced the service to five each way.

The winter timetable 1956 offered six each way plus one Saturdays, taking 29 minutes Berkeley Road–Lydney Town with a 3 minute call at Lydney Junction; the service in the opposite direction took 26 minutes with a 1 minute stop at Lydney Junction. In the summer of 1958 the Saturday train had been incorporated into the weekday service which offered seven trains each way.

During the autumn months of 1958-60, some passenger services between Birmingham, Snow Hill, Cheltenham and Swansea were diverted over the Severn Bridge to Lydney.

In October 1960 following the fall of the bridge, the weekday shuttle service of six trains each way daily

Sharpness-Berkeley Road continued, all calling at Berkeley. Despite the bridge's destruction, the 6th March to 2nd April, 1961 edition of Bradshaw's timetable hopefully listed trains running between Berkeley Road and Lydney Town.

The distance Sharpness-Lydney Junction was now 39 miles via Gloucester instead of 4¾ miles across the bridge. School trains ran Sharpness-Lydney via Gloucester until July 1962, then just three services each way Berkeley Road-Sharpness continuing until the passenger service was withdrawn on 7th September, 1964.

The Sunday Bristol-Cardiff and Portsmouth-Cardiff trains diverted via Gloucester extending the average journey time between the two cities to about 2½ hours rather than the usual 1 hour 10 minutes.

At least one Cardiff–Bournemouth excursion was run, the route being via the Severn Bridge, Mangotsfield, and Bath, an ex-GWR Class '63XX' 2-6-0 running through from Cardiff to Bath Green Park.

By May 1914 the GWR was working a 7.00 pm goods from Sharpness Docks-Gloucester via Lydney and when required a 11.30 pm Lydney Junction-Taunton via Berkeley Road Loop. In 1917 there were goods trains Sharpness-Stoke Gifford and Lydney Junction-Weston-super-Mare and Taunton with respective return workings, while in September 1917 a return working from Cardiff- Newton Abbot via the Severn Bridge was introduced.

Goods services were at their maximum in 1919 trains being Lydney Junction-Stoke Gifford and Taunton; Cardiff and Beachley-Stoke Gifford, (the Beachley portion with empty stone wagons en route to the MR's Thornbury branch. Its return working from Stoke Gifford-Beachley was double-headed to Sharpness where the train was divided, one portion taking stone to Beachley works, and the other the empties to Lydney Junction. Workings decreased after 1919, leaving only the MR's Sharpness-Gloucester and the GWR's Lydney Junction-Stoke Gifford until the bridge strike halted the latter in 1960.

BERKELEY ROAD JUNCTION. WORKING OF UP BRANCH TRAINS

The person conveying the Token from the Driver of an Up Branch train to the Signalman must first ascertain that the train, complete with tail lamp attached, has arrived on the loop line clear of the Single line and so advise the Signalman. See Table D.2.

BERKELEY

The points connecting the Siding with the Main Line are worked from a Ground Frame released by the Electric Staff for the Berkeley Road—Sharpness South Section or by the Electric Staff for the Berkeley Loop Junction—Sharpness South Section.

SHARPNESS

Oldminster Sidings

The position of the Signal (worked by Guards and Shunters from a Ground Frame) regulating the running of trains and engines on the Up Goods Siding fixed near the connections between the Sidings alongside the Up Goods Siding and the Goods Siding at Sharpness South Box is normally at " Clear ".

The Signal must be placed to " Danger " to protect shunting operations between the Sidings and the Down Goods Sidings or between the Up and Down Goods Sidings.

Truck Weighbridge Machine

Engines must not be allowed to pass over the truck weighbridge machine at Sharpness Station when the machine is in gear. No train or engine must pass over the truck weighbridge at a greater speed than 3 miles an hour.

North Dock Branch

A Ground Frame works the connection between the Single line and the North Dock Branch and is released by key on the Electric Token.

Only one engine may be allowed on the West side of the Docks at one and the same time, and when a train or engine has passed on to the North Docks Branch, no other train or engine may be allowed to go on to the Branch until the one from the West side has returned and the North Docks Branch is clear.

SWING BRIDGE. A fixed signal is provided at each end of the Viaduct to regulate the running of trains and engines over the Bridge; and catch points are provided to prevent any train or engine entering upon the Viaduct when the Bridge is open and the signals are at " Danger ". The signals and catch points are worked from the Ground Frames from which the Swing Bridge and the level crossing gates are controlled.

The catch points must always be kept in the throw-off position and the signals at " Danger ", except when it is necessary for a train or engine to pass, and before either of the signals is taken off, care must be taken to ascertain that the catch points are in the proper position, that the Swing Bridge is locked, and the line is clear, and that the level crossing gates are closed against the highway.

The Bridgeman will be responsible for working the bridge, points and signals, and for attending to the level crossing gates during the time he is on duty.

When the Bridgeman is not on duty, however, and a train is required to pass over the swing bridge, the Coal Tip Foreman or Shunter in charge of the work will be held responsible for working the points and signals, locking the bridge gates, etc.

Shunting at Sharpness Docks

It is important that Drivers shunting at Sharpness or running round the Docks should proceed cautiously, keep their engines under control, and be prepared to stop short of any obstruction ; the following instructions for regulating shunting at Sharpness must be strictly carried out :—

East Side.—Traffic may be exchanged between the Docks Sidings and the Commission's Sidings by either shunting engine but a proper understanding must first be reached by the two Foremen concerned before the shunt is made. A Stop lamp indicator is provided applicable to Down movements from Sharpness South to the Docks Sidings. The Commission's engines must not pass this stop lamp indicator until instructed by the Foreman.

West Side.—Neither the Commission's engine nor the Docks engine may pass the stop board marking the boundary between the Sidings unless accompanied by and in charge of the Foreman or Shunter on to whose Sidings the movement is to be made.

Vehicles with a wheel base exceeding 20 feet are prohibited in Sharpness Docks.

WORKING OF ENGINES OVER THE SEVERN BRIDGE

Trains running over the Severn Bridge must not be worked by more than one engine in front. Two engines coupled together must not, under any circumstances, be run over the Bridge.

In the event of the failure of an engine at either end of the Bridge and it is necessary for such engine to be taken to the opposite end, or if an engine fails on the Bridge, arrangements must be made for the engine to be worked specially, and four wagons must be placed between the assisting engine and the disabled engine. A competent man must in all cases ride upon the disabled engine.

136

Gloucester Traffic Department Working Timetable Appendix, October 1960.

Table 265 — **BERKELEY ROAD and LYDNEY (One class only)**

Severn and Wye—L.M.S. and G.W. Joint Line

Miles	Week Days only	a.m	a.m		p.m	p.m	p.m	
	Berkeley Road ..... dep	8 35	1145	..	2 25	4 50	6 35	..
2¼	Berkeley	8 41	1151	..	2 31	4 56	6 41	..
4	Sharpness	8 47	1157	..	2 37	5 2	6 47	..
5½	Severn Bridge D	8 52	12 4	..	2 42	5 7	6 52	..
8¼	Lydney Junction.. arr	8 59	1211	..	2 49	5 14	6 59	..
	dep	9 0	1212	..	2 50	5 15	7 0	..
8¾	Lydney Town....... arr	9 2	1214	..	2 52	5 17	7 2	..

Miles	Week Days only	a.m	a.m		p.m		p.m	p.m	
	Lydney Town...... dep	6 50	1025	..	1235	..	4 15	5 30	..
½	Lydney Junction.. arr	6 52	1028	..	1238	..	4 17	5 32	..
	dep	6 53	1029	..	1239	..	4 18	5 34	..
3¼	Severn Bridge D	7 0	1036	..	1246	..	4 25	5 41	..
4¾	Sharpness	7 7	1041	..	1251	..	4 31	5 46	..
6½	Berkeley................	7 13	1047	..	1257	..	4 37	5 52	..
8¾	Berkeley Road..... arr	7 19	1052	..	1 2	..	4 42	5 57	..

D Station for Blakeney.

October 1941 timetable.

Table 109 — **BERKELEY ROAD and LYDNEY TOWN**

WEEK DAYS ONLY—(Second class only)

Miles		am		am			am			pm E		pm S			pm S	pm S		pm E		pm		pm	
	Berkeley Road ... dep	8 22	..	1010	..	..	1152	..	..	1 55	..	2 15	..	..	..	4 45	..	5 6	..	6 28	..	8 50	..
2¼	Berkeley	8 28	....	1016	....	....	1158	....	....	2 1	....	2 20	....	....	....	4 50	....	5 12	....	6 34	....	8 55	....
4½	Sharpness	8 34	..	1022	..	..	12 4	..	..	2 7	..	2 27	..	..	..	4 57	..	5 18	..	6 40	..	9 1	..
5½	Severn Bridge for Blakeney	8 39	....	1027	....	....	12 9	....	....	2 12	....	2 32	....	....	....	5 2	....	5 23	....	6 45	....	9 6	....
8¼	Lydney Junction ... arr	3 46	..	1034	..	..	1216	..	..	2 19	..	2 39	..	..	..	5 9	..	5 30	..	6 52	..	9 14	..
	dep	8 49	....	1037	....	....	1219	....	....	2 21	....	2 42	....	....	4 55	5 12	....	5 33	....	6 55	....	9 17	....
9	Lydney Town ¶ .. arr	3 51	..	1039	..	..	1221	..	..	2 24	..	2 44	..	..	4 57	5 14	..	5 35	..	6 57	..	9 20	..

Miles		am			am			am			pm E		pm S	pm S	pm S			pm E	pm S	pm		pm	
	Lydney Town ... ... dep	7 10	..	..	9 30	..	..	1050	..	..	1 10	..	1 15	2 50	3 55	..	..	4 25	4 45	5 42	..	8 5	..
¾	Lydney Junction ... arr	7 12	....	....	9 32	....	....	1052	....	....	1 12	....	1 17	2 52	3 57	....	....	4 27	4 47	5 44	....	8 7	....
	dep	7 13	..	..	9 33	..	..	1053	..	..	1 13	..	1 18	..	3 58	..	..	4 28	..	5 45	..	8 8	..
3¼	Severn Bridge for Blakeney	7 20	....	....	9 40	....	....	11 0	....	....	1 20	....	1 25	....	4 5	....	....	4 35	....	5 52	....	8 15	....
4¾	Sharpness	7 25	..	..	9 45	..	..	11 5	..	..	1 25	..	1 30	..	4 10	..	..	4 40	..	5 57	..	8 20	..
6¾	Berkeley	7 30	....	....	9 50	....	....	1110	....	....	1 31	....	1 36	....	4 16	....	....	4 46	....	6 3	....	8 26	....
9	Berkeley Road .. arr	7 36	..	..	9 56	..	..	1116	..	..	1 36	..	1 41	..	4 21	..	..	4 51	..	6 8	..	8 31	..

E Except Saturdays. S Saturdays only.

¶ Bus Services operate between: Ross-on-Wye and Coleford; Coleford and Lydney; Coleford and Monmouth; and Coleford and Lydbrook

11th June, – 15th September, 1957 timetable.

BERKELEY ROAD AND SHARPNESS.

Miles.	Classification ..	Weekdays.						
			K		K			
			5.30 a.m. from Gloucester.		10.0 p.m. from Gloucester.			
	Target No. ..		18		16			
			a.m.		a.m.			
0	BERKELEY ROAD JUNC....dep.	..	6*50	..	11 50	..	..	..
	Berkeley Road Sth. Junc.	..	..	..	..	..	..	..
2¼	Berkeley	..	..	..	..	..	..	..
4¼	SHARPNESS................arr.	..	7 5	..	12 5	..	..	..

Miles.	Classification ..	Weekdays.							
				K		K			
				To Barnwood Sidings.		To Barnwood Sidings.			
	Target No. ..			18		16			
				a.m.	Berkeley arr. 12.47 p.m.	p.m.			
0	SHARPNESS ....dep.	..	..	7 55	..	12 40	..	..	..
2	Berkeley	..	..	..	..	12 52	..	..	..
	Berkeley Road Sth. Junc.	..	..	..	..	..	..	..	..
4¼	BERKELEY ROAD JUNC....arr.	..	..	8 10	..	1 2	..	..	..

Freight timetable 21st September, 1953, LMR operating area.

3

10.—Should an accident occur of such a nature as to obstruct the line, and the traffic is likely to be stopped for any considerable time, special arrangements must be made for working the trains to and from the staff station on each side of the obstruction until the line is again clear. The staff must be used to work trains between the obstruction and the staff station whence assistance can be obtained, and on the other side the traffic must be conducted by a pilotman, to be appointed by an order in writing (see following form), who must accompany every train between the obstruction and the staff station. The person in charge at each station in the section must have a copy of the pilotman's form. When the line is again clear no train must be allowed to pass the point where the obstruction existed, without both the staff and the pilotman. The pilotman must accompany the train to the staff station whence assistance was obtained, when the traffic must again be conducted according to these Regulations.

FORM REFERRED TO IN REGULATION 10.

GREAT WESTERN AND L. M. & S. RAILWAYS,
SEVERN AND WYE JOINT LINE.

WORKING OF SINGLE LINES OF RAILWAY BY ONLY ONE ENGINE IN STEAM OR TWO OR MORE ENGINES COUPLED TOGETHER.

WORKING BY PILOTMAN DURING OBSTRUCTION.

This Form must be filled up and used whenever it is temporarily necessary, owing to obstruction, to work the traffic by pilotman.

..............................*Station*

..............................19

The Single Line between..and......................................
being obstructed, the traffic between..and the place of obstruction will be worked by Pilotman in accordance with Regulation 10 of the Regulations for Working Single Lines of Railway by only one Engine in steam or two or more Engines coupled together.

....will act as Pilotman, and must accompany every train to and from..Station, and the point of obstruction.

This order is to remain in force until withdrawn by the Pilotman.

(*Signed*)..

To.. *Time*

Noted by..at.. ....
Noted by..at.. ....
Noted by..at.. ....
Noted by..at.. ....
Noted by..at place of obstruction..
Noted by..Pilotman.

The Pilotman must sign all the Forms issued and the Form retained by him must bear the signatures of every person to whom a copy of the Form is distributed.

Twelve of these forms must be kept in a convenient place at each end of the Line, so as to be available at any moment night or day.

A copy of this Form must be delivered to the Station Master or person in charge of the Staff Station where Pilot working commences, one must be retained by the Pilotman, and another must be conveyed by the Pilotman to the person in charge at the place of obstruction. If there is an intermediate Station, the person in charge must be supplied with a copy of the Form by the Pilotman at the first opportunity.

In the event of a Station Master himself acting as Pilotman, he must address and give a copy of the Form to the person he leaves in charge of his Station.

Station Masters and persons in charge receiving this Form will be held responsible that the Inspectors, Foremen, and others concerned at their Stations are immediately made acquainted with the circumstances, and are instructed in their necessary duties.

11.—Two competent men, provided with the necessary hand signals and detonators must be appointed to protect the obstruction, one on each side.

12.—All points on a single line that become facing points to trains running in either direction, if not interlocked, must either be padlocked or securely held by hand for the safe passage of trains.

Severn & Wye Joint Line Appendix to Service Timetables 11th April, 1932 (Pages 122-126).

6

ENGINE WHISTLES.

Standard Whistle Code—Applicable unless otherwise shown herein.

Main line	1 whistle.
Relief line	2 whistles.
To or from platform loops	2 whistles.
Branch lines	3 whistles.
Goods lines	4 whistles.
To engine sheds	2 short whistles.
Yards, to or from	4 short whistles.
Crossover road	Main line—1 crow and 1 whistle. Relief line—1 crow and 2 whistles.
In siding clear of running lines ...	3 short whistles.

SPECIAL STATION AND JUNCTION WHISTLES.

Lydney Docks over G.W. Line:—

East to Dock Sidings	3.
East to West side	2 and 1 crow.
Dock Sidings to West side	1 crow.
Dock Sidings to East side	1 long 1 short.
West to Docks	2.
Docks to West	2 short 1 long.

Drivers must sound their whistles when approaching the undermentioned crossings:—
Oldminster Level Crossing (Station side of Sharpness South Box).

Middle and Upper Forge Level Crossings between Lydney Town and Tufts Junction, between 9¼ and 10¼ mile posts.

Saw Mills Crossings, Coleford Branch, between Coleford Junction and Darkhill Sidings, between 13 and 13¼ mile posts. (Whistle boards provided.)

LIST OF PERMANENT SPEED RESTRICTIONS.

Name of place.	Between	Miles per hour.	Special instructions.
Sharpness	Goods Yard and Station Box	4	Drivers of up goods trains approaching Station Box must reduce speed to 4 miles per hour when passing through goods yard, and must exercise great care when working over goods line, and be prepared to stop at any point.
"	South Docks Junction to or from Passenger Station.	20	
"	Over North Dock Branch	6	
Severn Bridge	.. ..	..	No engine or train must cross the iron portion of the Severn Bridge in less than 3 minutes.
Lydney Junction	Otters Pool Junction to Lydney Junction Joint Station. Otters Pool Junction to Gt. Western Line	} 10	
Tufts Junction	Tufts Junction and Lydney Town ..	20	~~Up trains.~~ IN EACH DIRECTION. 1/3/37
" "	10m. 20ch. and 10m. 60ch.	20	In each direction.
Parkend	Over curves in up and down lines between 12¼ and 12½ mile posts.	10	
Coleford Branch	Coleford Junction to Coleford .. .. Coleford to Coleford Junction	} 15	Speed not to exceed 15 miles an hour at any point on this Branch.
" Junction	Double to single line	15	
Bicslade Siding	13m. 50ch. and 13m. 60ch.	20	In each direction.
Speech House Road Station	14m. 57ch. and 14m. 65ch.	15	(South End) in either direction.
" " "	14m. 70ch. and 14m. 77ch.	15	(North End) " "
Wimberry Branch Junction	15m. 12ch. and 15m. 45ch.	15	In either direction.
Serridge Junction	To and from the Lydbrook Branch ..	15	
Speculation Curve	16m. 57ch. and 17m. 0ch.	15	In either direction.
Waterloo Sidings	18m. 35ch. and 18m. 46ch.	15	(North End) in either direction.
Upper Lydbrook Station ..	18m. 60ch. and 18m. 76ch.	15	In either direction.
" " (South) ..	18m. 76ch. and 18m. 72ch.	10	In the direction of Serridge Junction.
" " " ..	18m. 72ch. and 18m. 60ch.	15	" " " "
" " (North) ..	19m. 5ch. and 19m. 10ch.	15	Through tunnel in either direction.
" " " ..	19m. 16ch. and 19m. 24ch.	15	In either direction.
Lydbrook Viaduct	19m. 66ch. and 19m. 76ch.	5	" "

Page 6—Add speed restrictions.

Between Berkeley Road and Sharpness—**40 miles per hour**—in each direction.

Between Severn Bridge and Otters Pool Junction—**25 miles per hour**—in each direction.

9

LOADING OF FREIGHT TRAINS WORKED BY G.W. ENGINES (exclusive of Brake Vans).

BRANCH.		For Group A Engines (G.W.). (Except where otherwise stated.) Number of wagons to be conveyed.			
From	To	Class 1 Traffic.	Class 2 Traffic.	Class 3 Traffic.	Empties.
Berkeley Rd. South Junc. ..	Sharpness South	36	43	54	60
Sharpness South	Lydney Junction	~~27~~ 25	~~32~~ 30	~~41~~ 38	~~54~~ 50
Lydney Junction	Coleford Junction	27.	32	41	54
Coleford Junction	Speech House Road ..	21	25	32	42
Speech House Road ..	Drybrook Road	10	12	15	~~25~~ 20
Drybrook Road	Speech House Road ..	35	42	53	60
Speech House Road ..	Coleford Junction	50 ~~45~~	54	60	60
Coleford Junction	Tufts Junction	50	54	60	60
Tufts Junction	Lydney Junction	50	54	60	60
Lydney Junction	Sharpness South	25 ~~27~~	~~32~~ 30	~~41~~ 38	~~54~~ 50
Sharpness South	Berkeley Rd. South Junc.	30	36	45	60
Tufts Junction	New Fancy Sidings ..	10	12	15	~~25~~ 20
New Fancy Sidings ..	Lightmoor Sidings ..	14	~~17~~ 14	~~21~~ 18	~~30~~ 25
Lightmoor Sidings	Foxes Bridge	22	26	33	44
Foxes Bridge	Drybrook Road	40	48	60	60
Drybrook Road	Foxes Bridge	14	17	21	30
Foxes Bridge	Lightmoor Sidings ..	35	42	53	60
Lightmoor Sidings	Moseley Green	50	54	60	60
Moseley Green	Tufts Junction	40	48	60	60
Drybrook Road	Cinderford	40	48	60	60
Cinderford	Drybrook Road	~~16~~ 12	~~16~~ 14	~~20~~ 18	~~28~~ 25
Tufts Junction	Princess Royal	9	11	14	~~22~~ 18
Princess Royal	Tufts Junction	50	54 ~~60~~	60	60
Coleford Junction	Coleford	7	8	11	14
Coleford	Milkwall	11	12	17	24
Milkwall	Coleford Junction	25	30	38	50
Serridge Junction	Miery Stock	20	24	30	40
Miery Stock	Lydbrook Junction ..	35	42	53	60
Lydbrook Junction ..	Miery Stock	~~12~~ 11	~~14~~ 13	~~18~~ 17.	~~28~~ 22.
Miery Stock	Serridge Junction	~~35~~ 40	42	53	60

BRANCH.		For "2251" class Engines. Number of wagons to be conveyed.			
From	To	Class 1 Traffic.	Class 2 Traffic.	Class 3 Traffic.	Empties.
Berkeley Road South ..	Sharpness South	42	50	63	70
Sharpness South	Sharpness Station	31	37	47	60
Sharpness Station	Lydney Junction	—	*	—	—
Sharpness South	Berkeley Road South ..	36	43	54	70

* Engines of "2251" class are prohibited on Severn Bridge.

For particulars of Maximum and Working Loads of G.W. Freight Trains running over the Joint Line between Lydney and Berkeley Road South Junction see G.W. Company's No. 7 Service Book.

INSTRUCTIONS FOR CALCULATING LOADS OF FREIGHT TRAINS WORKED BY G.W. ENGINES.

Loaded wagons will bear labels overprinted with the numerals 1 (coal, coke or patent fuel), 2 (other minerals), 3 (general merchandise), and guards, to arrive at the load of a train, must ascertain the number of wagons of each class of traffic, or empty wagons to be conveyed.

In order that due allowance may be made for certain heavy traffics in Classes 2 or 3, any wagons (except pitwood), although bearing Class 2 or 3 labels, *which are carrying contents weighing 7 tons or over, must be calculated as Class 1 for train loading.* Such wagons must, however, be entered on the guards' journals in the same columns as the number overprinted on the label.

Examples of traffics in Classes 2 or 3 which must be calculated as Class 1 are:—

Ballast.	Copper.	Ironstone.	Steel Bars.
Bricks.	Dolomite.	Pig Iron.	Tarmac.
Cement.	Granite.	Rails.	Tinplates.
China Clay.	Gravel.	Roadstone.	
China Stone.	Lime.	Sand.	

Pitwood bearing Class 3 labels must, for loading purposes, be calculated as Class 2 traffic.

The maximum and working loads apply (with few exceptions especially indicated) to ordinary freight wagons. Wagons of larger dimensions must be calculated as under:—

When Loaded.	When Empty.
Five 12 tonners = 6 Class 1	Three 15 tonners = 4 ordinary empty wagons
Two 15 " = 3 " 1	One 40 tonner = 3 ordinary empty wagons
Two 40 " = 7 " 1	8, 10 or 12 tonners to be regarded as ordinary empty wagons

20-ton wagons—See tables on page 10.

(*d*) **Lock indicators** in connection with each of the Three Electric Locks showing whether the locks are "**On,**" "**Off,**" or "**Wrong.**"

(*e*) **An electric switch** worked by lever No. 1 in the Swing Bridge Box, which, when the lever is in its normal position, prevents a tablet being withdrawn or the locks taken off the starting signals at each end of the Tablet Section.

(*f*) **An electric indicator** to show whether the Swing Bridge Pawl has fallen into proper position. This indicator shows "Pawl In" on a white disc and "Pawl Out" on a red disc.

(*g*) **Telephonic** communication between Severn Bridge Station Box, Swing Bridge Box, and Sharpness Station Box.

(*h*) **A bolting lever,** No. 2, in the Swing Bridge Box for bolting the Bridge.

(*j*) **Single needle through circuit,** each of the Boxes being on it.

(*k*) **A mechanical indicator** in the Swing Bridge Box, to show when the Swing Bridge is in position for bolting with lever No. 2.

Bell signals between Severn Bridge Station Box and Swing Bridge Box and between Swing Bridge Box and Sharpness Station Box.

See Clause.		Beats on Bell.	How to be given.
1	Call attention	1	
3	*Is line clear for passenger train?	4	3 pause 1.
3	*Is line clear for goods, mineral, or ballast train? ..	3	3 consecutively.
3	*Is line clear for light engine, or engine and brake? ..	5	2 pause 3.
3	Line clear	3	1 pause 2.
3	*Train arrived, ready to unlock bridge	3	2 pause 1.
3	*Train arrived, but another train waiting	5	3 pause 2.
3	Obstruction danger	6	6 consecutively.
4	Cancelling Is line clear signal	8	3 pause 5.
3	Release Bridge Lock	2	2 consecutively.
3	Bridge unlocked	4	2 pause 2.
11	*Testing bells	16	16 consecutively.
11	*Lock Bridge for testing	16	4 pause 4 pause 4 pause 4.
8	*Opening of signal box	15	5 pause 5 pause 5.
9	*Closing of signal box	17	7 pause 5 pause 5.
9	*Testing complete	16	8 pause 8.
12	*Testing controlled or slotted signals	20	5 pause 5 pause 5 pause 5.

Note.—These signals are to be given on the tapper bells.

Bell or Gong Signals between Severn Bridge Station Box and Sharpness Station Box.

See Clause.		Beats on Bell.	How to be given.
3	Release starting signal	7	3 pause 4.
3	Starting signal lowered	7	4 pause 3.

Note.—These signals are to be given on the train tablet instruments.

1.—**Call attention.**—The **Call attention** signal must always be given before the signals marked thus *, and must be acknowledged immediately on receipt.

2.—**Repetition and Acknowledgment of Signals.**—Except where special instructions are issued to the contrary, as in clause 3, no signal must be considered as understood until it has been correctly repeated to the signal box from which it was received. When the **Is line clear** signal is not promptly acknowledged it must given again at short intervals.

3.—**Mode of Signalling.**—When the signalman at either end of the tablet section has a train or engine approaching which has to cross the Bridge, he must first ascertain on the telephone if the signalman at the other end of the section is prepared to receive the train or engine, and, if so, he must call the attention of and give to the signalman at the Swing Bridge Box the proper **Is line clear** signal, and if the signalman at the Swing Bridge Box is in a position to close and lock the Bridge, he must, when the Bridge is in position, pull over lever No. 2, which will bolt the Bridge in the proper position for the Railway, and will itself be thereby automatically locked in this position. He must then pull over his **switch lever No. 1,** which will put the Severn Bridge Station Box and Sharpness Station Box in communication with each other on the Electric Train Tablet and Electric Lock Circuits and forward **Line clear** to the signalman in the rear and lower his signals for the train or engine to pass.

The signalman who has the train or engine waiting must then obtain a tablet in accordance with the Electric Train Tablet Block Regulations, and having done this, he must send the bell or gong signal **Release starting signal.**

The signalman receiving this signal must immediately hold down his special tapper-key to admit of the signal being lowered, and continue to do so until he receives the bell signal **Starting signal lowered.**

35

WORKING OF ENGINES OVER THE SEVERN BRIDGE.

Trains running over the Severn Bridge must not be worked by more than one engine in front. Two engines coupled together must not, under any circumstances, be run over the Bridge.

In the event of the failure of an engine at either end of the Bridge and its being necessary for such engine to be taken to the opposite end, or if an engine fail on the Bridge, arrangements must be made for the engine to be worked specially, and four wagons must be placed between the assisting engine and the disabled engine. A competent man must in all cases ride upon the disabled engine.

The following are the only engines that may be allowed to pass over the Severn Bridge—

GREAT WESTERN ENGINES.

0-6-0 class.—Nos. 363, 2301 to 2360, 2381 to 2490, 2511 to 2580.

0-6-0 (tank).—2021 to 2160.

L.M.S. FREIGHT ENGINES.

No. 1 class bearing numbers between 2399 and 2867.

No. 2 class bearing numbers between 2900 and 3129.

NOTE.—*Engines of other classes bearing numbers as above must not be allowed to pass over the Bridge.*

L.M.S. PASSENGER ENGINES.

No. 1 class (Precedent) type 2-4-0, straight link.—Nos. 5000, 5001, 5005, 5011, 5014, 5018, 5021, 5027, 5050, 5069, 5070.

No. 1 class (Waterloo) type 2-4-0, straight link.—No. 5095.

EXCHANGING PASSENGER VEHICLES AT LYDNEY JUNCTION.

The arrangements for exchanging passenger vehicles at Lydney Junction with the G.W. Company are as under:—

Vehicles from the Joint Line:—

Vehicles from the Joint Line to the Great Western Line off an up train must be left in the Sharpness road behind the Great Western Junction signal box, clear of all sidings, and the signalman at Otters Pool Joint signal box must advise the signalman in the Great Western Junction signal box by telephone.

Vehicles put off the down Joint passenger trains must be placed in the Otters Pool sidings on the Joint Line, and the signalman at Otters Pool must advise the signalman in charge of the Great Western Junction signal box.

If at any time the Joint engine is available, and the Great Western trains are waiting, it can be utilised to run any vehicle up to the Great Western platform to attach to the rear of a down train, but only under the direction of the Great Western foreman or shunter.

Vehicles from the Great Western Line:—

If the Joint engine is not available, vehicles for stations in the direction of Sharpness must be placed on the Sharpness road, immediately behind the Great Western Junction signal box, by the Great Western Company, but if the vehicles are for stations in the direction of Lydney Town and beyond, they must be placed in the Otters Pool siding, so that the engines of the Joint down trains may easily pick them up.

All messages exchanged in carrying out these instructions to be recorded and timed.

WORKING BETWEEN LYDNEY JUNCTION YARD AND LYDNEY DOCKS.

Use of Shunting Truck:—

A shunting truck is provided at Lydney Docks, and the following instructions must be observed:—

1.—In propelling wagons from Lydney Junction Yard to Lydney Basin, or Upper Dock to the sidings leading to Coal Tips Nos. 3 and 4, and also to sidings as far as Turntable Road or Long Siding, the man in charge of the shunting operations must walk in advance of the wagons.

2.—In propelling wagons from the Turntable Road or Long Siding to the Lower Docks or Harbour, the shunting truck must be used, and must be placed in front of the leading wagon for the accommodation of the men in charge of the shunting operations to ride upon, and then returned to the Turntable Road siding formed at the rear of train from Lower Dock.

No unauthorised person must be allowed to ride upon the shunting truck.

Western Region

British Railways Board

Transport Act 1962

Withdrawal of railway passenger services

The Minister of Transport has given his consent to the Board's proposal to discontinue all passenger train services between LYDNEY TOWN, SHARPNESS and BERKELEY ROAD and from the following stations:-

LYDNEY TOWN
SEVERN BRIDGE (for BLAKENEY)
SHARPNESS
BERKELEY

The terms of the Minister's consent can be inspected at local booking offices

The services will be withdrawn on and from Monday, 2nd November, 1964

Notice announcing the withdrawal of passenger services between Berkeley Road and Lydney Town.

'16XX' class 0-6-0PT No. 1627 passes Severn Bridge signal box working Up light engine 19th July, 1958. The signalman at the Midland Railway box stands ready to receive the tablet. *R. E. Toop*

The interior of Severn Bridge station signal box, 9th August, 1961. *Author*

Chapter Eleven

Signalling

1889 Section	**Type of Working**
Sharpness station – Sharpness East cabin	double line
Sharpness East cabin – Severn Bridge station	electric train tablet (Tyer's)
Severn Bridge station – Lydney Junction A cabin	electric train tablet
Lydney Junction A cabin – Lydney Town	double line

In 1903 the Sharpness Station and North boxes were amalgamated in one box on the Down platform.

In 1906 the block sections were	**Type of Working**
Berkeley Road – Berkeley Loop Junction	double line
Berkeley Loop Junction – Berkeley	double line
Berkeley – Sharpness South	double line
Sharpness South – Sharpness station	double line
Sharpness station – Severn Bridge station	train tablet No. 2 non-restoring type
Severn Bridge station – Lydney Junction (Otters Pool)	electric train tablet No. 2
Lydney Junction (Otters Pool) – Lydney engine shed	double line
Lydney engine shed – Lydney Town	double line

26th July, 1931 the three sections Berkeley Road – Sharpness South were singled, combined and controlled by electric train tablet, a separate electric train tablet system splitting the line into two sections when Berkeley Loop Junction box was switched in. Between Berkeley Road Junction and Berkeley Loop Junction, or Sharpness South, the person authorised to receive or deliver the electric token was the guard of a freight train or passenger train

Berkeley Road signal box is to the rear of the 2.15 pm Saturdays-only to Lydney Town.

R. E. Toop

Lydney Town: coal yard and builders' merchant siding on the right beyond the level crossing. Note the LMS signal.
Michael Wathen

with a guard, or fireman in the case of a light engine or passenger train without a guard.

For Sharpness South to Severn Bridge an auxiliary instrument was placed at the Severn Bridge end of the crossing loop at Sharpness North ground frame, a fireman obtaining permission by telephone to withdraw the electric token.

The Canal Swing Bridge Signal Box only had three signals, and one of those was for shipping. Its main purpose was to house the boilers, engines and controls of the swing bridge and the interlocking with the tablet machines at each end of the section.

The frame had seven Levers:
No. 1 Electric lock
No. 2 To lock and unlock bridge
No. 3 Spare
No. 4 Up Home signal
No. 5 Down Home signalling
No. 6 Spare
No. 7 Canal signal.

Contrary to normal practice, when a signalman at Sharpness or Severn Bridge had a train for the bridge, he was required to telephone to the other end of the section to ascertain if he could accept it. If he received an affirmative answer, he would send 'Call Attention' (single beat) to the Swing Bridge Box. When this was acknowledged he would send the appropriate 'Is Line Clear For ....?) bell code to the Swing Bridge.

Should the bridge be open for shipping, Swing Bridge would send 'Obstruction Danger'. If the bridge was closed but unlocked, the Swing Bridge man would reply by repeating the 'Is Line Clear For'. He would then pull Lever 2 followed by Lever 1. Lever 2 locked the bridge while Lever 1 locked Lever 2 and operated a switch which closed the electrical circuit between the tablet machines in the Sharpness and Severn Bridge boxes, thus allowing the release of the required tablet. The Swing Bridge man then cleared his appropriate Home signal: the Down signal in the approach cutting and the Up attached to the fixed span nearest the swing bridge. (As the fact that the bridge swung made the normal wire connection impossible, to operate the signal, the end of a sliding rod on the swing bridge made contact with the end of another sliding rod on land.)

Having received 'Train out of Section', Severn Bridge sent two beats to Sharpness and Severn Bridge requesting the Bridge Lock to be released. To do this both the Sharpness and Severn Bridge men had to press and hold the tablet machine bell plungers and tapper keys until receiving 2-2 from the Swing Bridge after he had normalised Levers 1 & 2. The requirement of three men to release the lock made accidental unlocking virtually impossible.

As the starting signal was released by the box at the opposite end of the section, that is Sharpness Down starter was released by the Severn Bridge box and vice versa, additional bell codes were given on the tablet machine bells, not the tapper bells:

Release Starting Signal 3 pause 4
Starting Signal Lowered 4 pause 3.

Other special bell codes used between the Severn Bridge, Swing Bridge and Sharpness boxes were:

Train arrived, ready to unlock bridge 2 pause 1
Train arrived, but another one waiting 3 pause 2
Release bridge lock 2 consecutively
Bridge unlocked 2 pause 2
Lock bridge for tests 4 pause 4 pause 4 pause 4

Originally the signal lamps were white for All Clear and red for Danger, but to conform with standard practice, in 1899 the white aspect was altered

A Railway Correspondence & Travel Society special approaches Berkeley Road South Junction en route to Sharpness, 26th September, 1959. Notice that the box is a GWR structure, but the signal LMS. The loop is double track, but the Berkeley Road – Lydney line on the right is single. The gentleman on the left is Colin Roberts mentioned in the acknowledgements.
Dr A. J. G. Dickens

west of Sharpness to green. The red distant lights were converted to yellow in 1945. Except for Sharpness station and Berkeley Loop Junction, the signal boxes were all of the MR type.

The GWR took over responsibility for signalling from 1st July, 1894 until 1st January, 1906 when it was passed to the MR, responsibility returning to the GWR in 1923. In 1948 responsibility changed from the London Midland Region of BR to the Western Region.

If a train from Lydney required to pass over the North Branch at Sharpness without going into the passenger station, it was permitted to do so and the guard responsible for taking the tablet from the driver to the signalman at Sharpness station box after seeing that the last vehicle of his train was clear of the main line. If a train was required to return over the North Branch towards Lydney without entering the passenger station, the driver had to send his fireman to the station signal box to collect the tablet.

Signals at the high-level swing bridge on the North Docks branch were required to be kept at Danger except when it was necessary for a train or engine to pass the catch points provided to prevent any train or engine entering the viaduct when the bridge was open. Both signals and catch points were worked from a ground frame, the bridgeman responsible for working the bridge, points and signals and attending to the level crossing gates. He was on duty 6.00 am to 5.30 pm from 1st March until 31st October, and 7.00 am to 5.00 pm 1st November to 28th February. When the bridgeman was not on duty and a train or engine required to pass, the coal tip foreman or shunter was responsible for carrying out his duty.

On 27th October, 1957 Sharpness Station box closed and responsibility for working the bridge from the east side was transferred to Sharpness South Junction signal box and at the same time the key token working in use from Berkeley Road to Sharpness was extended over the bridge.

On 9th May, 1965 one engine in steam working replaced the electric train token Berkeley Road Junction to Sharpness South signal box and the line designated a siding 13th October, 1968.

Hastings DMU 6L set No. 1017 near the low-level road/rail swing bridge on the Sharpness Dock internal rail system. The unit was working the Southern Electric Group/RCTS tour from Watford Junction on 12th October, 1985. It also visited Taunton, Portishead, Severn Beach and Tytherington. Note the signal on the left at the far end of the bridge.
Richard Giles

Chapter Twelve

Rebirth: The Vale of Berkeley Railway

On 17th October, 2015 the Vale of Berkeley Railway was inaugurated with the aim of re-opening Berkeley Road to Sharpness as a heritage line and possibly for commuter traffic as a housing consortium was proposing an eco-village of up to 5,000 houses in the Newtown and Berkeley area over the next 20 years. The Vale of Berkeley Railway was supported by a number of established preservationists including Andrew Goodman of the specialist rail vehicle road haulier Moveright International.

The railway is on good terms with Network Rail and Direct Rail Services: access was allowed to the Berkeley station site and Oldminster Sidings, the latter 300 yards in length, has been leased to the society for 25 years at a peppercorn rent. As Oldminster Sidings have been disused for many years, they were virtually in a forest of about 1,000 trees and bushes. Since 2017 these have been cleared and the four sidings are visible once more, but the vegetation keeps growing and a strimmer has to be used often. This is not the only problem as there is no ballast under the track, the sleepers are rotten, the water table is high and there is poor drainage. This will have to be dealt with before public trains can be run.

The Canal & River Trust's 40 year Masterplan for its land at Sharpness will see the use of the dock area diversified with new housing together with mixed-use sites. It has allowed the railway to set up a workshop in what was the former dock diesel locomotive shed. The Vale of Berkeley Railway Trust Engineering Services are capable of general milling and turning, including wheels up to 40 inches in diameter, gear cutting and surface grinding. This useful facility raises funds by offering engineering services in its comprehensive machine shop, thus generating income from outside organisations. Having outside work orders to fulfil during the Covid

Oldminster sidings, view Down 24th September, 2020. Hundreds of trees had been cleared from the site, as until 2017 it was wooded as on right.

Author

Midland Railway stop block, Oldminster sidings, 24th September, 2020.
Author

The Sharpness branch train staff in the ground frame, Berkeley, spring 2022.
Author's collection

epidemic meant that the railway was earning cash at a time it was closed to the public.

The Vale of Berkeley Railway is required to maintain a route to the docks should the resumption of freight traffic ever be required.

The immediate aim was to form a visitor centre at Oldminster Sidings and offer brake van rides. A bombshell fell in October 2020, when, following considerable work carried out planning the visitor centre, museum and train rides, it was discovered that large quantities of ammonium nitrate were stored at the docks rendering the sidings within an 'inner hazard zone', precluding public access and allowing the sidings only to be a restoration and maintenance facility only accessible by members.

The current plan for this slightly curved site is to retain Line 1 of the sidings as a running road complete with stop block. The remaining three sidings will be relaid to form two sidings each with a minimum of 120 metres of straight track at the north end of the site where a two-road locomotive shed will be built. The fourth siding will be truncated and converted to a rail vehicle loading and off-loading facility.

The Vale of Berkeley Railway and the Llanelli & Mynydd Mawr Railway have formed a useful partnership, their first project being in the spring of 2022 when the VoB Pacer No 143612 was sent to South Wales on a year's loan the VoB not immediately requiring it and thus ensuring it was kept in ready-to-run condition in addition to sharing maintenance and operating skills. The VoB's two Pacers Nos 143603 and 143612 are useful being compliant with all current main line requirements such as door interlocks and no drop windows.

After 7 years' work at the remains of Berkeley station, in 2023 the VoB was informed that as the nuclear waste disposal site comes within the licence of the Atomic Energy Authority the station site is now in a public exclusion zone. Hopes of purchasing the railway have been thwarted as it is intended to reuse the licensed atomic sites at Berkeley and Oldbury for small reactors so the line will be required for removing the nuclear waste probably into the next century. This problem has come on top of a background of covid, inflation and international conflict so the Vale of Berkeley Railway has not been able to develop as quickly as had been anticipated, but it has been enterprising in operating engineering and wood-working shops to earn cash.

The eco-village of up to 5,000 homes housing about 15,000 residents at Newtown, Sharpness, will require better public transport facilities so a case, part of the national *Restoring Your Railway* programme has been put forward for reopening the Berkeley Road-Sharpness branch to regular passenger traffic. Stroud District Council which has set a target towards sustainable, low carbon transport, is also proposing that the main line station be re-opened at Berkeley Road.

Vale of Berkeley Railway Locomotives

Number	Locomotive	Notes	Owner
15	Austerity 0-6-0ST built 1946 by Andrew Barclay for MoD	Wemyss Private Railway Loco Currently at Sharpness	Andrew Goodman
9642	'8750' class 0-6-0PT GWR built 1946 Swindon	At the flour mills under restoration	Andrew Goodman
4027	LMS Class '4F' 0-6-0 built 1924 Derby	Part of National Collection under restoration	National Railway Museum
D2069	'Class 03' 0-6-0DM built 1959 BR	Currently on loan to the Dean Forest Railway	Andrew Goodman
D9553	'Class 14' 0-6-0 built 1965 Swindon	In storage in Midlands	Andrew Goodman
7069	LMS diesel-electric shunter built 1935 English Electric	A forerunner of the BR 08 class	Mike Hoskin
2126	Fireless 0-4-0F built 1942 built Andrew Barclay	Donated to VoBR by Waterways Museum, Gloucester	VoBR Trust
-	Planet 4wDM built 1960	Worked at Dow Corning, Barry	Dr Andrew Woodhall
143 603	Class 143 'Pacer' built 1985	In storage at Wishaw, Warwickshire, but hopes to move to Sharpness	VoBR Trust
143 612	Class 143 'Pacer' built 1985	On loan to Llanelli & Mynydd Mawr Railway	VoBR Trust
W55025 *Pandora*	'Class 121' DMBS Bubble Car built 1960/1 by Pressed Steel Company	In storage	Andrew Goodman & Roland Hall

Network Rail Track Assessment Unit 950.001 leaving the Sharpness branch at Berkeley Road and proceeding towards Gloucester, 11th September, 2012. *Author*

No. 31285 heads the 3Q01 1754 Derby Research & Technical Centre-Exeter (Riverside) test working past Berkeley on 15th July, 2013. *Richard Giles*

A reproduction Lydney Town station on the Dean Forest Railway, 26th September, 2012. The line descends towards the Junction on a gradient of 1 in 261.
Author

'14XX' class 0-4-2T No. 1450 at Lydney Town working the 11.00 Parkend – Lydney Junction, 26th September, 2012.
Author

Suspension of services notice, at the Dean Forest Railway.
Author

Bibliography

Baxter, B., *British Locomotive Catalogue 1825-1923* Volume 3A (Moorland: Ashbourne, 1982)

Christiansen, R., *A Regional History of the Railways of Great Britain* Volume 13 Thames & Severn (David & Charles: Newton Abbot, 1981)

Berridge, P. S. A., *The Girder Bridge* (London: R. Maxwell, 1969)

Clinker, C. R., *Clinker's Register of Closed Passenger Stations and Goods Depots in England, Scotland and Wales 1830-1980* (Avon-Anglia Publications: Weston-super-Mare, 1988)

Cooke, R. A., *Track Layout Diagrams of the Great Western Railway and BR Western Region*, Section 20 (Author: Harwell, October 1988)

Cooke, R. A., *Track Layout Diagrams of the Great Western Railway and BR Western Region*, Section 37 (Author: Harwell, January 1996)

Dean Forest Railway Museum Trust, *Rails to the Forest* (Silver Link Publishing: Kettering, 2010)

Earl, J., & Hudson, S., *Midland Retrospective* (Midland Railway Society: Northampton, 2017)

Essery, R. J. & Jenkinson D., *An Illustrated Review of Midland Locomotives* Volume Three (Wild Swan: Didcot, 1988)

Fleming, D. J., *St Philip's Marsh* (D. Bradford Barton: Truro, 1982)

Gough, J., *The Midland Railway A Chronology* (Railway & Canal Historical Society: Mold, 1989)

Harrison, I., *Great Western Railway Locomotive Allocations for 1921* (Wild Swan: Didcot, 1984)

Huxley, R., *The Rise & Fall of the Severn Bridge Railway 1872-1970* (Amberley Publishing: Chalford, 2008)

Jordan, C., *Severn Enterprise* (Arthur Stockwell: Ilfracombe, 1977)

Lyons, E., *An Historical Survey of Great Western Engine Sheds 1947* (Oxford Publishing Company: Oxford, 1974)

Lewis, J., *Great Western Auto Trailers* Part Two (Wild Swan: Didcot, 1995)

Maggs, C. G., *The Swindon to Gloucester Line* (Alan Sutton: Stroud, 1991)

Maggs, C., *The Branch Lines of Gloucestershire* (Amberley Publishing: Stroud, 2011)

Maggs, C., *A History of the Great Western Railway* (Amberley Publishing: Stroud, 2013)

Maggs, C., *Great Britain's Railways* (Amberley Publishing: Stroud, 2018)

Mitchell V. & Smith K., *Branch Lines Around Lydney* (Middleton Press: Midhurst, 2008)

Oakley, M., *Gloucestershire Railway Stations* (The Dovecot Press: Wimborne, 2003)

Paar, H. W., *The Severn & Wye Railway* Part One (David & Charles: Newton Abbot, 1973)

Parkhouse, N., *Gloucester Midland Lines* Part 3 (Lightmoor Press: Lydney, 2019)

Pope, I., How, R. & Karau, P., *Severn & Wye Railway* Volume 1 (Wild Swan: Didcot, 1983)

Pocock, Revd N, & Harrison I., *Great Western Railway Locomotive Allocations for 1934* (Wild Swan: Didcot, 1987)

Railway Correspondence and Travel Society, *The Locomotives of the Great Western Railway*. (RCTS, various dates)

Smith, P., *An Historical Survey of the Midland in Gloucestershire* (Oxford Publishing Company: Oxford, 1985)

Weaver, C. P. & C. R., *The Gloucester and Sharpness Canal* (West Midlands Area Railway & Canal Historical Society: Wolverhampton, 1966)

Witts, C., *Severn Bridge Disaster* (River Severn Publications: Gloucester, 2010)

Newspapers & Magazines

Gloucester Journal
Engineering
Illustrated London News
Railway Magazine
Railway Observer
Railway World Vol 31 (1970) pp 446-451
Trains Illustrated

Appendix One

Industrial Locomotives

Lydney Tinplate Works						
Name	**Type**	**Builder**	**Works No**	**Year**	**Source**	**Disposal**
	4wDM	Ruston Hornsby	22163	1943	New	Austin Motor Co Ltd Co Cofton Hackett
Peter Pan	0-4-0ST	Hudswell, Clarke	831	1909	Morris Motors Ltd Coventry, 1946	Scrapped c.1953
Lydney	0-4-0ST	Andrew Barclay	1180	1911	Burry Old Works, Carms.1953	Scrapped 1958

Sharpness Docks						
Name	**Type**	**Builder**	**Works No**	**Year**	**Source**	**Disposal**
No. 1	0-4-0ST					Scrapped post 2/1919
No. 2	0-6-0ST	Sharp, Stewart	2356	1873	GWR No. 1398 4/1883	Scrapped 1924
SD3	0-4-0ST	Avonside	1446	1902	Withdrawn 1964	Scrapped?
No. 4	0-4-0ST	Peckett	812	1900	Ex-War Dept, 1920	Sold after 1947
SD 5	0-4-0ST	Avonside	1906	1924	New	Scrapped 1962
No. 6	0-4-0ST	Peckett	2060	1944	New	Scrapped 1962
DL1 *Reg*	0-4-0DM	Ruston Hornsby	463150	1961	New	Sold 11/1998
DL2	0-6-0DM	Bagnall	3151	1962	New	Sold 11/1998

Sharpness Docks DL2, a Bagnall 0-6-0 diesel-mechanical of 1962, 6th May, 1975. *Revd. Alan Newman*

Ex-BR Locomotives Scrapped by Cooper's at Sharpness 1964-5		
Class	**Type**	**Number**
'County'	4-6-0	1006, 1027
'14XX'	0-4-2T	1409
'2251'	0-6-0	2245
'28XX'	2-8-0	2842, 2872
'8750'	0-6-0PT	3633
'Castle'	4-6-0	4087, 5040, 5050, 5071, 7015*
'Hall'	4-6-0	4924, 4996, 5905, 5935, 5943, 5954, 5978
'42XX'	2-8-0T	5262
'54XX'	0-6-0PT	5420
'58XX'	0-4-2T	5810
'43XX'	2-6-0	6319, 6365, 7335

*Sold to Cashmore, Newport, but scrapped at Sharpness.

Delivered to Cooper, Sharpness, but later sold to Cashmore, Newport, where they were scrapped:

Class	Type	Number
'Castle'	4-6-0	7009, 7037
'Hall'	4-6-0	5986
'47XX'	2-8-0	4701

Severn & Wye 0-6-0T *Little John* as MR No. 1123A. Built by Fletcher, Jennings in 1874 it was withdrawn January 1905. *Author's collection*

Appendix Three

Severn & Wye Railway Locomotives transferred to the GWR and MR in 1895							
To GWR							
Name	**Type**	**Builder**	**Date**	**Works No.**	**GWR No.**	**Valuation £**	**Withdrawn**
Maid Marian	0-6-0T	Avonside	1872	940	1357	860	March 1910
Will Scarlet	0-6-0T	Fletcher, Jennings	1873	122	1356	200	September 1923
Alan-a-Dale	0-6-0T	Fletcher, Jennings	1876	157	1355	300	January 1905
Ranger	0-6-0ST	Avonside*		–	1358	500	November 1897
Wye	0-4-0T	Fletcher, Jennings	1876	153	1359	300	December 1910
Severn Bridge	0-6-0T	Vulcan	1880	860	1354	1160	February 1906
Gaveller	0-6-0T	Vulcan	1891	1309	1353	1300	November 1903

* A 0-6-0 probably Hawthorn Works No. 708 built for the London & North Western Railway 1848/9, rebuilt to saddle-tank by Avonside January 1891.

To MR							
Name	**Type**	**Builder**	**Date**	**Works No.**	**MR No.**	**Valuation £**	**Withdrawn**
Robin Hood	0-6-0T	Fletcher, Jennings	1872	83	1121A	300	*c.*1896
Friar Tuck	0-6-0T	Avonside	1872	810	1122A	500	November 1911
Little John	0-6-0T	Fletcher, Jennings	1874	140	1123A	1100	January 1905
Sharpness	0-6-0T	Vulcan	1880	859	1124A	860	January 1924
Sabrina	0-6-0T	Vulcan	1882	953	1125A	860	August 1920
Forester	0-6-0T	Vulcan	1886	1163	1126A	1000	December 1924

Appendix Four

Severn & Wye Railway Passenger Stock transferred to GWR & MR in 1895		
Type	**GWR**	**MR**
First/Third-class compo	Nos. 1, 5, 11, 13, 16	Nos. 2, 10, 14
First/Third-class brake compo	–	6, 8
Brake Third	3, 7, 9	4
Third-class	15, 17	–
Brake van	18	–
Horse box	19	–
Carriage truck	20	–
First-class saloon	–	12

All were 4-wheeled, painted light brown and valued at £120 for No. 12, the others only being worth £12 – £25.

Appendix Five

Accounts

Expenditure	£	s	d	January – June 1894 £	s	d	Corresponding Period 1893 £
To Maintenance							
Way, Works & Stations	3029	17	11				3148
Harbour & docks	629	14	11				455
Coal tips & cranes	50	16	5				32
				3710	9	3	3635
Loco power, including							
Stationary engines				3524	16	5	3451
Carriage & wagon repairs				248	18	8	437
Traffic expenses (coaching& merchandise)				2710	12	7	2389
General charges				871	14	9	1007
Rates & taxes				966	5	9	632
Government duty				129	13	8	103
Compensation (accidents & losses)					15	0	79
Law charges				59	18	6	26
Miscellaneous expenses				95	3	4	122
Travelling expenses				137	8	2	125
Advertising, postages, etc				129	5	3	118
				12,585	1	4	12,124
Chancery expenses				200	0	0	
Road & Canal Traffic Act				43			
Reserve fund for bad debts etc				110	16	8	50
Balance, carried to net revenue account				7714	15	5	5287
				20,610	13	5	17,504

Credit	Receipts £	s	d	£	s	d	£
Passengers	2184	10	2				2083
Parcels, horses, carriages, etc	350	5	6				259
Mails	40	0	0				40
				2574	15	8	2382
Minerals & merchandise				16,331	18	7	13675
Wharfage				861	10	8	730
Dock & harbour dues				505	0	2	385
Rents				328	3	4	326
Transfer fees				9	5	0	6
				20,610	17	1	17,504

Index